Lecture Notes of
the Unione Matematica Italiana

More information about this series at http://www.springer.com/series/7172

Editorial Board

Unione
Matematica
Italiana

The Editorial Policy can be found
at the back of the volume.

Edoardo Ballico • Alessandra Bernardi •
Iacopo Carusotto • Sonia Mazzucchi • Valter Moretti
Editors

Quantum Physics and Geometry

 Springer

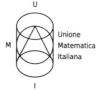 Unione Matematica Italiana

Editors

Edoardo Ballico
Dipartimento di Matematica
Università di Trento
Trento, Italy

Alessandra Bernardi
Dipartimento di Matematica
Università di Trento
Trento, Italy

Iacopo Carusotto
BEC Center
INO-CNR
Trento, Italy

Sonia Mazzucchi
Dipartimento di Matematica
Università di Trento
Trento, Italy

Valter Moretti
Dipartimento di Matematica
Università di Trento
Trento, Italy

ISSN 1862-9113   ISSN 1862-9121   (electronic)
Lecture Notes of the Unione Matematica Italiana
ISBN 978-3-030-06121-0   ISBN 978-3-030-06122-7   (eBook)
https://doi.org/10.1007/978-3-030-06122-7

Library of Congress Control Number: 2019932799

Mathematics Subject Classification (2010): Primary: 81-XX, Secondary: 14-XX

This Springer imprint is published by the registered company Springer Nature Switzerland AG
The registered company address is: Gewerbestrasse 11, 6330 Cham, Switzerland

Contents

Contributors

Edoardo Ballico Dipartimento di Matematica, Università di Trento, Trento, Italy

Alessandra Bernardi Dipartimento di Matematica, Università di Trento, Trento, Italy

Iacopo Carusotto BEC Center, INO-CNR, Trento, Italy

Luca Chiantini Dipartimento di Ingegneria dell'Informazione e Scienze Matematiche, Università di Siena, Siena, Italy

F. M. Ciaglia Sezione INFN di Napoli and Dipartimento di Fisica E. Pancini dell'Universitá Federico II di Napoli, Complesso Universitario di Monte S. Angelo, Naples, Italy

Frédéric Holweck Laboratoire Interdisciplinaire Carnot de Bourgogne, University Bourgogne Franche-Comté, Belfort, France

A. Ibort ICMAT, Instituto de Ciencias Matemáticas (CSIC-UAM-UC3M-UCM) and Depto. de Matemáticas, Univ. Carlos III de Madrid, Leganés, Madrid, Spain

Joseph M. Landsberg Department of Mathematics, Texas A&M University, College Station, TX, USA

G. Marmo Sezione INFN di Napoli and Dipartimento di Fisica E. Pancini dell'Universitá Federico II di Napoli, Complesso Universitario di Monte S. Angelo, Naples, Italy

Sonia Mazzucchi Dipartimento di Matematica, Università di Trento, Trento, Italy

Valter Moretti Dipartimento di Matematica, Università di Trento, Trento, Italy

Davide Pastorello Department of Mathematics, University of Trento, Trento Institute for Fundamental Physics and Applications (TIFPA), Povo, Trento, Italy

Bassano Vacchini Dipartimento di Fisica "Aldo Pontremoli", Università degli Studi di Milano, Milan, Italy

INFN, Sezione di Milano, Milan, Italy

Chapter 1
Introduction

**Edoardo Ballico, Alessandra Bernardi, Iacopo Carusotto,
Sonia Mazzucchi, and Valter Moretti**

The development of quantum mechanics has been one of the greatest scientific achievements of the early twentieth century. In spite of its remarkable success in explaining and predicting an amazing number of properties of our physical world, its interpretation has raised strong controversies among a wide community of scientists and philosophers. One of the hottest points of discussion is the meaning of the so-called quantum entanglement that, for systems of two or many particles, allows in particular the possibility for each particle of the system to be simultaneously located at different spatial positions. Entangled states display a special kind of correlations. Generally speaking, differently from the statistical correlations that are usually found in classical probability theory, quantum entanglement cannot be understood in terms of statistically distributed hidden variables and must involve the possibility for quantum systems of particles to be simultaneously in different single particle pure quantum states. Entangled states therefore present facets of the quantum worlds which are even more complicated than the famous example of a superposition of states in the so-called Schrödinger's cat which is simultaneously classically dead and alive. The peculiar phenomenology of quantum mechanics goes far beyond this paradoxical case: in contrast to the usual chain rules of classical conditional probability, the probability for a physical event to occur in a quantum framework

E. Ballico · A. Bernardi (✉)
Dipartimento di Matematica, Università di Trento, Trento, Italy
e-mail: edoardo.ballico@unitn.it; alessandra.bernardi@unitn.it

I. Carusotto
BEC Center, INO-CNR, Trento, Italy
e-mail: iacopo.carusotto@unitn.it

S. Mazzucchi · V. Moretti
Dipartimento di Matematica, Università di Trento, Trento, Italy
e-mail: sonia.mazzucchi@unitn.it; valter.moretti@unitn.it

© Springer Nature Switzerland AG 2019   1
E. Ballico et al. (eds.), *Quantum Physics and Geometry*,
Lecture Notes of the Unione Matematica Italiana 25,
https://doi.org/10.1007/978-3-030-06122-7_1

is computed by the interference of the complex-valued amplitudes corresponding to the different classical states. In dynamical processes, these classical positional states are described by paths that the system can follow during its evolution. This description of the physical world is commonly known as Feynman integral and implicitly requires that the system be simultaneously in different classical states at all intermediate times [1]. The mathematical counterpart of this picture is that quantum states of a composite system are described by a tensor product structure where each product entry represents a component of the system. In this picture, entanglement is encoded in quantum superpositions, that is linear combinations of completely decomposed tensors. In this sense, if the tensor product involves different states of a given component which are localized in far and causally separated spatial regions, a single component of the system may be simultaneously located in different places.

While the observable consequences of quantum mechanics have been experimentally explored all along the twentieth century, starting from the discrete energy levels of the hydrogen atom towards superconductivity and superfluidity in quantum condensed matter physics and precision measurements in quantum relativistic particle physics, the most basic and profound features of entanglement and its philosophical consequences have started being investigated only much more recently. A crucial step in this development was the formulation in 1935 of the so-called Einstein-Podolsky-Rosen (EPR) paradox raising doubts on the completeness of the quantum mechanical description of the physical world [2] in view of the existence of entangled states in the formalism of quantum theory and the Luders-von Neumann postulate on the instantaneous collapse of the wavefunction after a measurement procedure. The subsequent derivation in 1964 of the so-called Bell inequalities [3] was the milestone, which offered a quantitative criterion to test quantum mechanics against alternative hidden variable theories satisfying a local realism principle and essentially ruling out entangled states as proposed in the EPR paper. So far, the outcome of all experiments carried out along these lines starting from Aspect's 1982 one on cascaded photon emission [4] has been a strong confirmation of the predictions of quantum mechanics predicting violation of Bell's inequalities and ruling out the local realism principle. In the following years, the experiments have been gradually improved to better deal with various hidden assumptions or loopholes pointed out by various scientists. In 2015, for the first time, the violation of Bell's inequalities was corroborated by an experimental test of Bell's theorem by R. Hanson et al. certifying the absence of any additional assumptions or loophole [5].

In addition to a revolution in our philosophical understanding of the physical world around us, the success of quantum mechanics in describing these amazing features of the microscopic world has then given a dramatic boost into the exploration of their possible use in technological applications, e.g. to the quantum communication and quantum information processing, two new branches of science based on a dramatic change in perspective in logics and computation. As one can easily imagine, this paradigm shift is accompanied by the need of new mathematical and computer science tools for the description and the control of quantum mechanical systems and, more practically, for the full exploitation of the new possibilities opened by entanglement for communication and computation.

This special volume was prepared in the wake of the "International workshop on Quantum Physics and Geometry" organized during July 2017 in Levico Terme (Trento, Italy) (http://www.science.unitn.it/~carusott/QUANTUMGEO17/index.html) on these topics. This event, sponsored by CIRM with the precious support of INDAM, University of Trento, TIFPA-INFN and the INO-CNR BEC Center gathered world specialists in both physical sciences and in mathematics, with the aim of exploring possible interdisciplinary links between quantum information and geometry and contributing to the creation of a community of researchers trying to export advanced mathematical concepts to this new applicative field. The objective was to convey to a single event leading experts from the two fields, so to explore interdisciplinary connections and contribute establishing an active and long-lasting community. On the physics side, a conductive thread of the event has been the characterization of entanglement; on the mathematics one, different tools to describe it from different perspectives have been covered, including tensor decomposition, the classification of the orbit closures of some Lie groups, tensor network representations, and topological properties of the quantum states. The articles that follow give a hint of the rich developments that one may expect to result from this meeting of different worlds. While all contributions present exciting state-of-the-art results, they are also meant to offer a general, mathematics-oriented introduction to quantum science and technologies and to their latest developments.

The first contribution by J.M. Landsberg on "A very brief introduction to quantum computing and quantum information theory for mathematicians" summarizes the PhD course on "Quantum Information and Geometry" that he has given at Trento University with the support of INDAM during the months of June and July 2017 surrounding the Levico workshop. In combination with the recorded lectures that are available under request (https://drive.google.com/open?id=0B2Y1CpIKbFuSR1hVT3BfNmtTSFU), this long article aims at giving a complete coverage of the background material from both physics and computer science. The contribution by D. Pastorello on "Entanglement, CP-maps and quantum communications" reviews basic concepts of quantum mechanics and entanglement and then focuses on the potential of quantum entanglement as a resource in communication systems. The contribution by B. Vacchini on "Frontiers of open quantum system dynamics" presents important developments on the dynamics of quantum systems coupled to environments, which generalize to a wider context the quantum evolution in terms of the well-known Schrödinger equation. Mathematical results on the use of advanced geometrical concepts in quantum information theory are presented in the contribution by F. Holweck on "Geometric constructions over $\mathbb{C}$ and $\mathbb{F}^2$ for Quantum Information", with a special attention to the entanglement of pure multipartite systems and to contextuality issues [6]. In both problems, a central role is played by representation theory, which is respectively used to classify entanglement in terms of the closure diagram of the orbits in tensor spaces and for the description of commutation relations of the generalized N-qubit Pauli group. The contribution by L. Chiantini on "Hilbert functions and tensor analysis" illustrates the power of geometric methods for the decomposition of tensors and, in particular, offers a survey-style introduction to the important problem of the uniqueness of the decomposition (the so called "identifiability"), useful for signal processing and, possibly, for the representation of quantum states of many indistinguishable particles. As a final point, some extension to the famous Kruskal's criterion is proposed. Finally, the contribution by M. Ciaglia, A. Ibort and G. Marmo on "Differential Geometry of Quantum States, Observables and Evolution" summarizes an alternative geometric description of quantum mechanical systems in terms of the Kähler geometry of the space of pure states of a closed quantum system and discusses how the composition of systems and the resulting entanglement can be captured in this new setting.

We hope that this volume will trigger an active interest from the mathematical community towards the exciting challenges that quantum science and technology is raising to scientists of all disciplines.

References

1. R.P. Feynman, *QED: The Strange Theory of Light and Matter* (Princeton University Press, Princeton, 2006)
2. A. Einstein, B. Podolsky, N. Rosen, Can quantum-mechanical description of physical reality be considered complete? Phys. Rev. **47**, 777 (1935)
3. J.S. Bell, *Speakable and Unspeakable in Quantum Mechanics* (Cambridge University Press, Cambridge, 1987)
4. A. Aspect, P. Grangier, G. Roger, Experimental realization of Einstein-Podolsky-Rosen-Bohm Gedankenexperiment: a new violation of Bell's inequalities. Phys. Rev. Lett. **49**, 91 (1982)
5. R. Hanson et al., Loophole-free Bell inequality violation using electron spins separated by 1.3 kilometers. Nature **526**, 682 (2015)
6. S. Kochen, E.P. Specker, The problem of hidden variables in quantum mechanics, in *The Logico-Algebraic Approach to Quantum Mechanics* (Springer, Dordrecht, 1975), pp. 293–328

Chapter 2
A Very Brief Introduction to Quantum Computing and Quantum Information Theory for Mathematicians

Joseph M. Landsberg

Abstract This is a very brief introduction to quantum computing and quantum information theory, primarily aimed at geometers. Beyond basic definitions and examples, I emphasize aspects of interest to geometers, especially connections with asymptotic representation theory. Proofs can be found in standard references such as Kitaev et al. (Classical and quantum computation, vol. 47. American Mathematical Society, Providence, 2002) and Nielson and Chuang (Quantum computation and quantum information. Cambridge University Press, Cambridge, 2000) as well as Landsberg (Quantum computation and information: Notes for fall 2017 TAMU class, 2017).

2.1 Overview

I begin, in Sect. 2.2, by presenting the postulates of quantum mechanics as a natural generalization of probability theory. In Sect. 2.3 I describe basic entanglement phenomena of "super dense coding", "teleportation", and Bell's confirmation of the "paradox" proposed by Einstein-Podolsky-Rosen. In Sect. 2.4 I outline aspects of the basic quantum algorithms, emphasizing the geometry involved. Section 2.5 is a detour into classical information theory, which is the basis of its quantum cousin briefly discussed in Sect. 2.7. Before that, in Sect. 2.6, I reformulate quantum theory in terms of density operators, which facilitates the discussion of quantum information theory. Critical to quantum information theory is *von Neumann entropy* and in Sect. 2.8 I elaborate on some of its properties. A generalization of "teleportation" is discussed in Sect. 2.9. Regarding practical computation, the exponential growth in size of $(\mathbb{C}^2)^{\otimes n}$ with n that appears in quantum information theory leads to the notion of "feasible" states discussed in Sect. 2.10, which has interesting algebraic geometry associated to it. I conclude with a discussion of representation-theoretic

J. M. Landsberg (✉)
Department of Mathematics, Texas A&M University, College Station, TX, USA
e-mail: jml@math.tamu.edu

© Springer Nature Switzerland AG 2019
E. Ballico et al. (eds.), *Quantum Physics and Geometry*,
Lecture Notes of the Unione Matematica Italiana 25,
https://doi.org/10.1007/978-3-030-06122-7_2

aspects of quantum information theory, including a discussion of the quantum marginal problem in Sect. 2.11. I do not discuss topological quantum computing, which utilizes the representation theory of the braid group. For those interested in more details from this perspective, see [18].

2.2 Quantum Computation as Generalized Probabilistic Computation

In this section I take the point of view advocated in [1] and other places that quantum computing should be viewed as a natural generalization of probabilistic computing, and more generally that the laws of quantum mechanics as generalizations of the laws of probability.

2.2.1 Classical and Probabilistic Computing via Linear Algebra

This section is inspired by Arora and Barak [2, Exercise 10.4].

Classical communication deals with *bits*, elements of $\{0, 1\}$, which will be convenient to think of as elements of $\mathbb{F}_2$, the field with two elements. Let $f_n : \mathbb{F}_2^n \to \mathbb{F}_2$ be a sequence of functions. Give $\mathbb{R}^2$ basis $\{|0\rangle, |1\rangle\}$ (such notation is standard in quantum mechanics) and give $(\mathbb{R}^2)^{\otimes m} = \mathbb{R}^{2^m}$ basis $\{|I\rangle \mid I \in \{0, 1\}^m\}$. In this way, we may identify $\mathbb{F}_2^m$ with the set of basis vectors of $\mathbb{R}^{2^m}$. A computation of f_n (via an arithmetic or Boolean circuit) may be phrased as a sequence of linear maps on a vector space containing $\mathbb{R}^{2^n}$, where each linear map comes from a pre-fixed set agreed upon in advance. In anticipation of what will come in quantum computation, the pre-fixed set of maps (called *gates* in the literature) will be taken from maps having the following properties:

1. Each linear map must take probability distributions to probability distributions. This implies the matrices are *stochastic*: the entries are non-negative and each column sums to 1.
2. Each linear map only alters a small number of entries. For simplicity assume it alters at most three entries, i.e., it acts on at most $\mathbb{R}^{2^3}$ and is the identity on all other factors in the tensor product.

In quantum computation, the first property will be replaced by requiring the linear maps to be completely positive and trace preserving (see Sect. 2.7). The second is the same and justified because "universal" quantum computing is possible with such maps, even requiring the three factors to be adjacent, which is essentially due to the classical Cartan-Dieudonné theorem.

To facilitate comparison with quantum computation, first restrict to reversible classical computation. The complexity class of a sequence of functions in classical reversible computation is the same as in arbitrary classical computation.

For example, if we want to effect $(x, y) \mapsto x * y$, consider the map

$$|x, y, z\rangle \mapsto |x, y, z \oplus (x * y)\rangle = |x, y, z \oplus (x \wedge y)\rangle \tag{2.1}$$

(where the second expression is for those preferring Boolean notation) and act as the identity on all other basis vectors (sometimes called *registers*). Here z will represent "workspace bits": x, y will come from the input and z will always be set to 0 in the input. In the basis $|000\rangle, |001\rangle, |010\rangle, |100\rangle, |011\rangle, |101\rangle, |110\rangle, |111\rangle$, of $\mathbb{R}^8$, the matrix is

$$\begin{pmatrix} 1 & 0 & 0 & 0 & 0 & 0 & 0 & 0 \\ 0 & 1 & 0 & 0 & 0 & 0 & 0 & 0 \\ 0 & 0 & 1 & 0 & 0 & 0 & 0 & 0 \\ 0 & 0 & 0 & 1 & 0 & 0 & 0 & 0 \\ 0 & 0 & 0 & 0 & 1 & 0 & 0 & 0 \\ 0 & 0 & 0 & 0 & 0 & 1 & 0 & 0 \\ 0 & 0 & 0 & 0 & 0 & 0 & 0 & 1 \\ 0 & 0 & 0 & 0 & 0 & 0 & 1 & 0 \end{pmatrix}. \tag{2.2}$$

This gate is sometimes called the *Toffoli gate* and the matrix the *Toffoli matrix*.

The swap (negation) gate $\neg$ is realized by the matrix

$$\sigma_x = \begin{pmatrix} 0 & 1 \\ 1 & 0 \end{pmatrix}. \tag{2.3}$$

The swap and Toffoli matrices can perform any computation that is accomplished via a sequence of matrices drawn from some finite set of Boolean operations, each acting on a fixed number of basis vectors with at worst a polynomial in n size increase in the number of matrices needed. For those familiar with Boolean circuits, any sequence of Boolean circuits (one for each n) may be replaced by a sequence with just Toffoli and negation gates with at worst a polynomial (in n) blow up in size.

A probability distribution on $\{0, 1\}^m$ may be encoded as a vector in $\mathbb{R}^{2^m}$: If the probability distribution assigns probability p_I to $I \in \{0, 1\}^m$, assign to the distribution the vector $v = \sum_I p_I |I\rangle \in \mathbb{R}^{2^m}$.

The matrices (2.2), (2.3) realize classical computation. To add randomness to enable probabilistic computation, introduce the matrix

$$\begin{pmatrix} \frac{1}{2} & \frac{1}{2} \\ \frac{1}{2} & \frac{1}{2} \end{pmatrix}$$

which acts on a single $\mathbb{R}^2$ corresponding to a fair coin flip. Note that the coin flip matrix is not invertible, which will be one motivation for quantum computation in Sect. 2.2.2. Work in $\mathbb{R}^{2^{n+s+r}}$ where r is the number of times one needs to access a random choice and s is the number of matrices (arithmetic operations) in addition to the coin tosses needed to compute f.

A probabilistic computation, viewed this way, starts with $|x0^{r+s}\rangle$, where $x \in \mathbb{F}_2^n$ is the input. One then applies a sequence of admissible stochastic linear maps to it, and ends with a vector that encodes a probability distribution on $\{0, 1\}^{n+s+r}$. One then restricts this to $\{0, 1\}^{p(n)}$, that is, one takes the vector and throws away all but the first $p(n)$ entries. This vector encodes a probability sub-distribution, i.e., all coefficients are non-negative and they sum to a number between zero and one. One then renormalizes (dividing each entry by the sum of the entries) to obtain a vector encoding a probability distribution on $\{0, 1\}^{p(n)}$ and then outputs the answer according to this distribution. Note that even if our calculation is feasible (i.e., polynomial in size), to write out the original output vector that one truncates would be exponential in cost. A stronger variant of this phenomenon will occur with quantum computing, where the result will be obtained with a polynomial size calculation, but one does not have access to the vector created, even using an exponential amount of computation.

To further prepare for the analogy with quantum computation, define a probabilistic bit (a *pbit*) to be an element of

$$\{p_0|0\rangle + p_1|1\rangle \mid p_j \in [0, 1] \text{ and } p_0 + p_1 = 1\} \subset \mathbb{R}^2.$$

Note that the set of pbits (possible states) is a convex set, and the basis vectors are the extremal points of this convex set.

2.2.2 A Wish List

Here is a wish list for how one might want to improve upon the above set-up:

1. Allow more general kinds of linear maps to get more computing power, while keeping the maps easy to compute.
2. Have reversible computation: we saw that classical computation can be made reversible, but the coin flip was not. This property is motivated by physics, where many physical theories require time reversibility.
3. Again motivated by physics, one would like to have a continuous evolution of the probability vector, more precisely, one would like the probability vector to depend on a continuous parameter t such that if $|\psi_{t_1}\rangle = X|\psi_{t_0}\rangle$, then there exist admissible matrices Y, Z such that $|\psi_{t_0 + \frac{1}{2}t_1}\rangle = Y|\psi_{t_0}\rangle$ and $|\psi_{t_1}\rangle = Z|\psi_{t_0 + \frac{1}{2}t_1}\rangle$ and $X = ZY$. In particular, one wants operators to have square roots. (Physicists sometimes state this as "time evolution being described by a semi-group".)

One way to make the coin flip reversible is, instead of making the probability distribution be determined by the sum of the coefficients, one could take the sum of the squares. If one does this, there is no harm in allowing the entries of the output vectors to become negative, and one could use

$$H := \frac{1}{\sqrt{2}} \begin{pmatrix} 1 & 1 \\ 1 & -1 \end{pmatrix} \tag{2.4}$$

for the coin flip. The matrix H is called the *Hadamard matrix* or *Hadamard gate* in the quantum computing literature. If we make this change, we obtain our second wish, and moreover have many operations be "continuous", because the set of matrices preserving the norm-squared of a real-valued vector is the *orthogonal group* $O(n) = \{A \in Mat_{n \times n} \mid AA^T = \mathrm{Id}\}$. So for example, any rotation has a square root.

However our third property will not be completely satisfied, as the matrix

$$\begin{pmatrix} 1 & 0 \\ 0 & -1 \end{pmatrix}$$

which represents a reflection, does not have a square root in $O(2)$.

To have the third wish satisfied, allow vectors with *complex* entries. From now on let $i = \sqrt{-1}$. For a complex number $z = x + iy$ let $\bar{z} = x - iy$ denote its complex conjugate and $|z|^2 = z\bar{z}$ the square of its norm.

So we go from pbits, $\{p|0\rangle + q|1\rangle \mid p, q \geq 0$ and $p + q = 1\}$ to *qubits*, the set of which is

$$\{\alpha|0\rangle + \beta|1\rangle \mid \alpha, \beta \in \mathbb{C} \text{ and } |\alpha|^2 + |\beta|^2 = 1\}. \tag{2.5}$$

The set of qubits, considered in terms of real parameters, looks at first like the 3-sphere S^3 in $\mathbb{R}^4 \simeq \mathbb{C}^2$. However, the probability distributions induced by $|\psi\rangle$ and $e^{i\theta}|\psi\rangle$ are the same so it is really S^3/S^1 (the Hopf fibration), i.e., the two-sphere S^2. In the quantum literature this is referred to as the *Bloch sphere*. Geometrically, it would be more natural (especially since we have already seen the need to re-normalize in probabilistic computation) to work with projective space $\mathbb{CP}^1 \simeq S^2$ as our space of qubits, instead of a subset of $\mathbb{C}^2$. So the set of qubits is better seen as (2.5) modulo the equivalence $|\psi\rangle \sim e^{i\theta}|\psi\rangle$.

For $v = (v_1, \ldots, v_n) \in \mathbb{C}^n$, write $|v|^2 = |v_1|^2 + \cdots + |v_n|^2$. The set of stochastic matrices is now replaced by the unitary group

$$\mathbf{U}(n) := \{A \in Mat_{n \times n}(\mathbb{C}) \mid |Av| = |v| \ \forall |v\rangle \in \mathbb{C}^n\}.$$

The unitary group satisfies the third wish on the list: For all $A \in \mathbf{U}(n)$, there exists a matrix $B \in \mathbf{U}(n)$ satisfying $B^2 = A$.

Consider wish 1: it is an open question! However at least our generalized probabilistic computation includes our old probabilistic computation because the matrices (2.2), (2.3), (2.4) are unitary.

An indication that generalized probability may be related to quantum mechanics is that the interference patterns observed in the famous two slit experiments is manifested in generalized probability: one obtains a "random bit" by applying H to $|0\rangle$: $H|0\rangle = \frac{1}{\sqrt{2}}(|0\rangle + |1\rangle)$. However, if one applies a second quantum coin flip, one loses the randomness as $H^2 = \mathrm{Id}$ so $H^2|0\rangle = |0\rangle$, which, as pointed out in [1], could be interpreted as a manifestation of interference.

2.2.3   Postulates of Quantum Mechanics and Relevant Linear Algebra

Here are the standard postulates of quantum mechanics and relevant definitions from linear algebra.

P1 Associated to any isolated physical system is a Hilbert space $\mathcal{H}$, called the *state space*. The system is completely described at a given moment by a unit vector $|\psi\rangle \in \mathcal{H}$, called its *state vector*, which is well defined up to a phase $e^{i\theta}$ with $\theta \in \mathbb{R}$. Alternatively one may work in projective space $\mathbb{P}\mathcal{H}$.

Explanations A *Hilbert space* $\mathcal{H}$ is a (complete) complex vector space endowed with a non-degenerate Hermitian inner-product, $h : \mathcal{H} \times \mathcal{H} \to \mathbb{C}$, where by definition h is linear in the first factor and conjugate linear in the second, $h(|v\rangle, |w\rangle) = \overline{h(|w\rangle, |v\rangle)}$ for all v, w, and $h(|v\rangle, |v\rangle) > 0$ for all $|v\rangle \neq 0$.

The Hermitian inner-product h allows an identification of $\mathcal{H}$ with $\mathcal{H}^*$ by $|w\rangle \mapsto \langle w| := h(\cdot, |w\rangle)$. This identification will be used repeatedly. Write $h(|v\rangle, |w\rangle) = \langle w|v\rangle$ and $|v| = \sqrt{\langle v|v\rangle}$ for the *length* of $|v\rangle$.

If $\mathcal{H} = \mathbb{C}^n$ with its standard basis, where $|v\rangle = (v_1, \ldots, v_n)$, the *standard Hermitian inner-product* on $\mathbb{C}^n$ is $\langle w|v\rangle = \sum_{j=1}^{n} \overline{w}_j v_j$. I will always assume $\mathbb{C}^n$ is equipped with its standard Hermitian inner-product.

Remark 2.2.1 When studying quantum mechanics in general, one needs to allow infinite dimensional Hilbert spaces, but in the case of quantum computing, one restricts to finite dimensional Hilbert spaces, usually $(\mathbb{C}^2)^{\otimes N}$.

P2 The state of an isolated system evolves with time according to the *Schrödinger equation*

$$i\hbar \frac{d|\psi\rangle}{dt} = X|\psi\rangle$$

where $\hbar$ is a constant (*Planck's constant*) and X is a fixed *Hermitian operator*, called the *Hamiltonian* of the system. (Physicists, enamored of the letter H, often also use it to denote the Hamiltonian.) Here, recall that the *adjoint* of an operator $X \in \text{End}(\mathcal{H})$, is the operator $X^\dagger \in \text{End}(\mathcal{H})$ such that $\langle X^\dagger v | w \rangle = \langle v | Xw \rangle$ for all $v, w \in \mathcal{H}$ and X is *Hermitian* if $X = X^\dagger$. For a general Hilbert space, the Unitary group is $\mathbf{U}(\mathcal{H}) := \{ U \in \text{End}(\mathcal{H}) \mid |Uv| = |v| \; \forall | v \rangle \in \mathcal{H} \}$.

How is generalized probability related to Schrödinger's equation? Let $U(t) \subset \mathbf{U}(\mathcal{H})$ be a smooth curve with $U(0) = \text{Id}$. Write $U'(0) = \frac{d}{dt}|_{t=0} U(t)$. Consider

$$
\begin{aligned}
0 &= \frac{d}{dt}\Big|_{t=0} \langle v | w \rangle \\
&= \frac{d}{dt}\Big|_{t=0} \langle U(t)v | U(t)w \rangle \\
&= \langle U'(0)v | w \rangle + \langle v | U'(0)w \rangle.
\end{aligned}
$$

Thus $iU'(0)$ is Hermitian. We are almost at Schrödinger's equation. Let $\mathfrak{u}(\mathcal{H}) = T_{\text{Id}}\mathbf{U}(\mathcal{H})$ denote the Lie algebra of $\mathbf{U}(\mathcal{H})$ so $i\mathfrak{u}(\mathcal{H})$ is the space of Hermitian endomorphisms. For $X \in \text{End}(\mathcal{H})$, write $X^k \in \text{End}(\mathcal{H})$ for $X \cdots X$ applied k times. Write $e^X := \sum_{k=0}^{\infty} \frac{1}{k!} X^k$. If X is Hermitian, then $e^{iX} \in \mathbf{U}(\mathcal{H})$. Postulate 2 implies the system will evolve unitarily, by (assuming one starts at $t = 0$), $| \psi_t \rangle = U(t) | \psi_0 \rangle$, where

$$
U(t) = e^{\frac{-itX}{\hbar}} .
$$

Measurements Our first two postulates dealt with isolated systems. In reality, no system is isolated and the whole universe is modeled by one enormous Hilbert space. In practice, parts of the system are sufficiently isolated that they can be treated as isolated systems. However, they are occasionally acted upon by the outside world, and one needs a way to describe this outside interference. For our purposes, the isolated systems will be the Hilbert space attached to the input in a quantum algorithm and the outside interference will be the measurement at the end. That is, after a sequence of unitary operations one obtains a vector $| \psi \rangle = \sum z_j | j \rangle$ (here implicitly assuming the Hilbert space is of countable dimension), and as in generalized probability:

P3 The probability of obtaining outcome j under a measurement is $|z_j|^2$.

In Sect. 2.6, motivated again by probability, **P1, P3** will be generalized to new postulates that give rise to the same theory, but are more convenient to work with in information theory.

A typical situation in quantum mechanics and quantum computing is that there are two or more isolated systems, say $\mathcal{H}_A, \mathcal{H}_B$ that are brought together (i.e., allowed to interact with each other) to form a larger isolated system $\mathcal{H}_{AB}$. The larger system is called the *composite system*. In classical probability, the composite space is $\{0, 1\}^{N_A} \times \{0, 1\}^{N_B}$. In our generalized probability, the composite space is $(\mathbb{C}^2)^{\otimes N_A} \otimes (\mathbb{C}^2)^{\otimes N_B} = (\mathbb{C}^2)^{\otimes (N_A + N_B)}$:

P4 The state of a composite system $\mathcal{H}_{AB}$ is the tensor product of the state spaces of the component physical systems $\mathcal{H}_A$, $\mathcal{H}_B$: $\mathcal{H}_{AB} = \mathcal{H}_A \otimes \mathcal{H}_B$.

When dealing with composite systems, we will need to allow partial measurements whose outcomes are of the form $|I\rangle \otimes |\phi\rangle$ with $|\phi\rangle$ arbitrary.

This tensor product structure gives rise to the notion of *entanglement*, which accounts for phenomenon outside of our classical intuition, as discussed in the next section.

Definition 2.2.2 A state $|\psi\rangle \in \mathcal{H}_1 \otimes \cdots \otimes \mathcal{H}_n$ is called *separable* if it corresponds to a rank one tensor, i.e., $|\psi\rangle = |v_1\rangle \otimes \cdots \otimes |v_n\rangle$ with each $|v_j\rangle \in \mathcal{H}_j$. Otherwise it is *entangled*.

2.3 Entanglement Phenomena

2.3.1 Super-Dense Coding[1]

Physicists describe their experiments in terms of two characters, Alice and Bob. I generally follow this convention. Let $\mathcal{H} = \mathbb{C}^2 \otimes \mathbb{C}^2 = \mathcal{H}_A \otimes \mathcal{H}_B$, and let $|epr\rangle = \frac{|00\rangle + |11\rangle}{\sqrt{2}}$ (called the *EPR state* in the physics literature after Einstein-Podolsky-Rosen). Assume this state has been created, both Alice and Bob are aware of it, Alice is in possession of the first qubit, and Bob the second. In particular Alice can act on the first qubit by unitary matrices and Bob can act on the second. This all happens before the experiment begins.

Now say Alice wants to transmit a two classical bit message to Bob, i.e., one of the four states $|00\rangle$, $|01\rangle$, $|10\rangle$, $|11\rangle$ by transmitting qubits. We will see that she can do so transmitting just one qubit. If she manipulates her qubit by acting on the first $\mathbb{C}^2$ by a unitary transformation, $|epr\rangle$ will be manipulated. She uses the following matrices depending on the message she wants to transmit:

to transmit	act by	to obtain
$|00\rangle$	Id	$\frac{|00\rangle + |11\rangle}{\sqrt{2}}$
$|01\rangle$	$\begin{pmatrix} 1 & 0 \\ 0 & -1 \end{pmatrix} =: \sigma_z$	$\frac{|00\rangle - |11\rangle}{\sqrt{2}}$
$|10\rangle$	$\begin{pmatrix} 0 & 1 \\ 1 & 0 \end{pmatrix} =: \sigma_x$	$\frac{|10\rangle + |01\rangle}{\sqrt{2}}$
$|11\rangle$	$\begin{pmatrix} 0 & -1 \\ 1 & 0 \end{pmatrix} =: -i\sigma_y$	$\frac{|01\rangle - |10\rangle}{\sqrt{2}}$

---

[1]Physicists use the word "super" in the same way American teenagers use the word "like".

where the names $\sigma_x, \sigma_y, \sigma_z$ are traditional in the physics literature (the *Pauli matrices*). If Alice sends Bob her qubit, so he is now in possession of the modified $|epr\rangle$ (although he does not see it), he can determine which of the four messages she sent him by measuring the state in his possession. More precisely, first Bob acts on $\mathbb{C}^2 \otimes \mathbb{C}^2$ by a unitary transformation that takes the orthonormal basis in the "to obtain" column to the standard orthonormal basis (this is a composition of two Hadamard matrices), to obtain a state vector whose probability is concentrated at one of the four classical states. He then measures, and obtains the correct classical state with probability one.

In summary, with preparation of an EPR state in advance, plus transmission of a single qubit, one can transmit two classical bits of information.

2.3.2 Quantum Teleportation

Here again, Alice and Bob share half of an EPR state, Alice is in possession of a qubit $|\psi\rangle = \alpha|0\rangle + \beta|1\rangle$, and wants to "send" $|\psi\rangle$ to Bob. However Alice is only allowed to transmit classical information to Bob. We will see that she can accomplish her goal by transmitting two classical bits. Write the state of the system as

$$\frac{1}{\sqrt{2}} [\alpha|0\rangle \otimes (|00\rangle + |11\rangle) + \beta|1\rangle \otimes (|00\rangle + |11\rangle)]$$

where Alice can operate on the first two qubits. If Alice acts on the first two qubits by $H \otimes \sigma_x = \frac{1}{\sqrt{2}} \begin{pmatrix} 1 & 1 \\ 1 & -1 \end{pmatrix} \otimes \begin{pmatrix} 0 & 1 \\ 1 & 0 \end{pmatrix}$, she obtains

$$\frac{1}{2} [|00\rangle \otimes (\alpha|0\rangle + \beta|1\rangle) + |01\rangle \otimes (\alpha|1\rangle + \beta|0\rangle) + |10\rangle \otimes (\alpha|0\rangle - \beta|1\rangle) + |11\rangle \otimes (\alpha|1\rangle - \beta|0\rangle)] .$$

Notice that Bob's coefficient of Alice's $|00\rangle$ is the state $|\psi\rangle$ that is to be transmitted. Alice performs a measurement. If she has the good luck to obtain $|00\rangle$, then she knows Bob has $|\psi\rangle$ and she can tell him classically that he is in possession of $|\psi\rangle$. But say she obtains the state $|01\rangle$: the situation is still good, she knows Bob is in possession of a state such that, if he acts on it with $\sigma_x = \begin{pmatrix} 0 & 1 \\ 1 & 0 \end{pmatrix}$, he will obtain the state $|\psi\rangle$, so she just needs to tell him classically to apply σ_x. Since they had communicated the algorithm in the past, all Alice really needs to tell Bob in the first case is the classical message 00 and in the second case the message 01. The cases of 10 and 11 are similar.

In summary, a shared EPR pair plus sending two classical bits of information allows transmission of one qubit.

Remark 2.3.1 In the literature this phenomenon is named *quantum teleportation*. Since information is transmitted at a speed slower than the speed of light, the use of the word "teleportation", which implies instantaneous transmission, is misleading.

2.3.3 Bell's Game

The 1935 Einstein-Podolsky-Rosen paper [10] challenged quantum mechanics with the following thought experiment that they believed implied instantaneous communication across distances, in violation of principles of relativity: Alice and Bob prepare $|epr\rangle = \frac{1}{\sqrt{2}}(|00\rangle + |11\rangle)$, then travel far apart. Alice measures her bit. If she gets 0, then she can predict with certainty that Bob will get 0 in his measurement, even if his measurement is taken a second later and they are a light year apart.

Ironically, this thought experiment has been made into an actual experiment. One modern interpretation (see, e.g., [2]) is that there is no paradox because the system does not transmit information faster than the speed of light, but rather they are acting on information that has already been shared. What follows is a version from [7], adapted from the presentation in [2].

Charlie chooses $x, y \in \{0, 1\}$ at random and sends x to Alice and y to Bob. Based on this information, Alice and Bob, without communicating with each other, get to choose bits a, b and send them to Charlie. The game is such that Alice and Bob play on a team. They win if $a \oplus b = x \wedge y$, i.e., either $(x, y) \neq (1, 1)$ and $a = b$ or $(x, y) = (1, 1)$ and $a \neq b$.

2.3.3.1 Classical Version

Note that if Alice and Bob both always choose 0, they win with probability $\frac{3}{4}$.

Theorem 2.3.2 ([3]) *Regardless of the strategy Alice and Bob use, they never win with probability greater than* $\frac{3}{4}$.

See, e.g., [2, Thm. 10.3] for a proof.

2.3.3.2 Quantum Version

Although there is still no communication allowed between Alice and Bob, they will exploit a pre-shared $|epr\rangle$ to gain an advantage over the classical case. Alice and Bob prepare $|epr\rangle = \frac{|00\rangle + |11\rangle}{\sqrt{2}}$ in advance, and Alice takes the first qubit and Bob the second. When Alice gets x from Charlie, if $x = 1$, she applies a rotation by $\frac{\pi}{8}$ to her qubit, and if $x = 0$ she does nothing. When Bob gets y from Charlie, he applies a rotation by $-\frac{\pi}{8}$ to his qubit if $y = 1$ and if $y = 0$ he does nothing. (The order these rotations are applied does not matter because the corresponding operators on

$(\mathbb{C}^2)^{\otimes 2}$ commute.) Both of them measure their respective qubits and send the values obtained to Charlie.

Theorem 2.3.3 *With this strategy, Alice and Bob win with probability at least $\frac{4}{5}$.*

The idea behind the strategy is that when $(x, y) \neq (1, 1)$, the states of the two qubits will have an angle at most $\frac{\pi}{8}$ between them, but when $(x, y) = (1, 1)$, the angle will be $\frac{\pi}{4}$. That is, when $(x, y) \neq (1, 1)$, the manipulation makes it more likely that Alice and Bob's measurements produce the same outcomes, and less likely to produce the same outcome when $(x, y) = (1, 1)$. See [2, Thm. 10.4] for details.

2.4   Quantum Algorithms

Rather than giving a detailed description of the algorithms, I just present a few main ideas that illustrate the differences with classical and probabilistic algorithms.

2.4.1   Grover's Search Algorithm

The problem: given $F_n : \mathbb{F}_2^n \to \mathbb{F}_2$, computable by a $poly(n)$-size classical circuit, find a such that $F_n(a) = 1$ if such a exists.

Grover found a quantum circuit of size $poly(n)2^{\frac{n}{2}}$ that solves this problem with high probability. Compare this with a brute force search, which requires a circuit of size $poly(n)2^n$. No classical or probabilistic algorithm is known that does better than $poly(n)2^n$. Note that it also gives a size $poly(n)2^{\frac{n}{2}}$ probabilistic solution to the **NP**-complete problem SAT (it is stronger, as it not only determines existence of a solution, but finds it).

I present the algorithm for the following simplified version where one is promised there exists exactly one solution. All essential ideas of the general case are here.

Problem Given $F_n : \mathbb{F}_2^n \to \mathbb{F}_2$, computable by a $poly(n)$-size classical circuit, and the information that there is exactly one vector a with $F_n(a) = 1$, find a.

The idea of the algorithm is to start with a vector equidistant from all possible solutions, and then to incrementally rotate it towards a. What is strange for our classical intuition is that one is able to rotate towards the solution without knowing what it is, and similarly, we won't "see" the rotation matrix either.

Work in $(\mathbb{C}^2)^{\otimes n+s}$ where $s = s(n)$ is the size of the classical circuit needed to compute F_n. I suppress reference to the s "workspace bits" in what follows.

The following vector is the average of all the classical (observable) states:

$$|av\rangle := \frac{1}{2^{\frac{n}{2}}} \sum_{I \in \{0,1\}^n} |I\rangle. \tag{2.6}$$

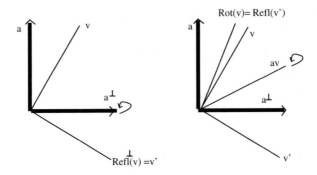

Fig. 2.1 Rotation of v to $Rot(v)$ via two reflections

To prepare $|av\rangle$, note that $H|0\rangle = \frac{1}{\sqrt{2}}(|0\rangle + |1\rangle)$, so applying $H^{\otimes n}$ to $|0\cdots 0\rangle$ transforms it to $|av\rangle$. The cost of this is n gates (matrices).

Since $|av\rangle$ is equidistant from all possible solution vectors, $\langle av|a\rangle = \frac{1}{2^{\frac{n}{2}}}$. We want to rotate $|av\rangle$ towards the unknown a. Recall that $\cos(\angle(|v\rangle, |w\rangle)) = \frac{\langle v|w\rangle}{|v||w|}$. Write the angle between av and a as $\frac{\pi}{2} - \theta$, so $\sin(\theta) = \frac{1}{2^{\frac{n}{2}}}$.

A rotation is a product of two reflections. In order to perform the rotation Rot that moves $|av\rangle$ towards $|a\rangle$, first reflect in the hyperplane orthogonal to $|a\rangle$, and then in the hyperplane orthogonal to $|av\rangle$, as in Fig. 2.1, which is valid for rotating any vector $|v\rangle$ towards $|a\rangle$.

Consider the map

$$|xy\rangle \mapsto |x(y \oplus F(x))\rangle \tag{2.7}$$

defined on basis vectors and extended linearly. To execute this, use the s workspace bits that are suppressed from the notation, to effect s reversible classical gates. Initially set $y = 0$ so that the image is $|x0\rangle$ for $x \neq a$, and $|x1\rangle$ when $x = a$. Next apply the quantum gate $\mathrm{Id} \otimes \begin{pmatrix} 1 & 0 \\ 0 & -1 \end{pmatrix}$ which sends $|x0\rangle \mapsto |x0\rangle$, and $|x1\rangle \mapsto -|x1\rangle$. Finally apply the map $|xy\rangle \mapsto |x(y \oplus F(x))\rangle$ again.

Thus $|a0\rangle \mapsto -|a0\rangle$ and all other vectors $|b0\rangle$ are mapped to themselves, as desired.

Next we need to reflect around $|av\rangle$. It is easy to reflect around a classical state, so first perform the map $H^{-1\otimes n} = H^{\otimes n}$ that sends $|av\rangle$ to $|0\cdots 0\rangle$, then reflect in the hyperplane perpendicular to $|0\cdots 0\rangle$ using the Boolean function $g : \mathbb{F}_2^n \to \mathbb{F}_2$ that outputs 1 if and only if its input is $(0, \ldots, 0)$, in the role of F for our previous reflection, then apply $H^{\otimes n}$ again so the resulting reflection is about $|av\rangle$. (Note that both these reflections have polynomial size cost.)

The composition of these two reflections is the desired rotation Rot. The vector $Rot|av\rangle$ is not useful as measuring it only slightly increases the probability of obtaining $|a\rangle$, but if one composes Rot with itself $O(\frac{1}{\theta})$ times, one obtains a vector much closer to $|a\rangle$. (Note that $\theta \sim \sin(\theta)$ so $\frac{1}{\theta} \sim \sqrt{N}$.)

For more details, see, e.g., [2, Thm. 10.13] or [20, §6.1].

2.4.2 The Quantum Discrete Fourier Transform

Underlying the famous quantum algorithm of Shor for factoring integers and Simon's algorithm that led up to it, are "quantum" versions of the discrete Fourier transform on finite abelian groups.

The DFT for $\mathbb{Z}/M\mathbb{Z}$, in vector notation, for $j \in \mathbb{Z}/M\mathbb{Z}$, is

$$|j\rangle \mapsto \frac{1}{\sqrt{M}} \sum_{k=0}^{M-1} \omega^{jk} |k\rangle$$

where $\omega = e^{\frac{2\pi i}{M}}$. It is a unitary change of basis such that in the new basis, multiplication in $\mathbb{Z}/M\mathbb{Z}$ is given by a diagonal matrix, and the classical FFT writes the DFT as a product of $O(\log(M))$ sparse matrices (each with $M << M^2$ nonzero entries), for a total cost of $O(\log(M)M) < O(M^2)$ arithmetic operations.

Say $M = 2^m$. The DFT can be written as a product of $O(m^3) = O(\log(M)^3)$ controlled local unitary operators. Hence one can approximately construct the output vector by a sequence of $poly(m)$ unitary operators from our gate set with the caveat that we won't be able to "see" it.

Here is the quantum DFT: It will be convenient to express j in binary and view $\mathbb{C}^M = (\mathbb{C}^2)^{\otimes m}$, i.e., write

$$|j\rangle = |j_1\rangle \otimes \cdots \otimes |j_m\rangle$$

where $j = j_1 2^{m-1} + j_2 2^{m-2} + \cdots + j_m 2^0$ and $j_i \in \{0, 1\}$. Write the DFT as

$$|j_1\rangle \otimes \cdots \otimes |j_m\rangle$$

$$\mapsto \frac{1}{\sqrt{M}} \sum_{k=0}^{M-1} \omega^{jk} |k\rangle$$

$$= \frac{1}{\sqrt{M}} \sum_{k_i \in \{0,1\}} \omega^{j(\sum_{l=1}^{m} k_l 2^{m-l})} |k_1\rangle \otimes \cdots \otimes |k_m\rangle$$

$$= \frac{1}{\sqrt{M}} \sum_{k_i \in \{0,1\}} \bigotimes_{l=1}^{m} \left[\omega^{jk_l 2^{m-l}} |k_l\rangle \right]$$

$$= \frac{1}{\sqrt{M}} \sum_{k_i \in \{0,1\}} \bigotimes_{l=1}^{m} \left[\omega^{(j_1 2^{2m-1-l} + \cdots + j_m 2^{m-l})k_l} |k_l\rangle \right]$$

$$= \frac{1}{2^{\frac{m}{2}}} (|0\rangle + \omega^{j_m 2^{-1}} |1\rangle) \otimes (|0\rangle + \omega^{j_{m-1} 2^{-1} + j_m 2^{-2}} |1\rangle) \otimes (|0\rangle + \omega^{j_{m-2} 2^{-1} + j_{m-1} 2^{-2} + j_m 2^{-3}} |1\rangle)$$

$$\tag{2.8}$$

$$\otimes \cdots \otimes (|0\rangle + \omega^{\sum_{s=0}^{m-1} j_{m-s} 2^{m-(s+1)}} |1\rangle)$$

where for the last line if $2m - s - l > m$, i.e., $s + l < m$, there is no contribution with j_s because $\omega^{2^m} = 1$, and I multiplied all terms by $1 = \omega^{-2^m}$ to have negative exponents.

It will be notationally more convenient to write the quantum circuit for this vector with the order of factors reversed, so I describe a quantum circuit that produces

$$\frac{1}{\sqrt{2}} (|0\rangle + \omega^{\sum_{s=0}^{m-1} j_{m-s} 2^{m-(s+1)}} |1\rangle) \otimes \cdots \otimes \frac{1}{\sqrt{2}} (|0\rangle + \omega^{j_{m-2} 2^{-1} + j_{m-1} 2^{-2} + j_m 2^{-3}} |1\rangle)$$

$$\tag{2.9}$$

$$\otimes \frac{1}{\sqrt{2}} (|0\rangle + \omega^{j_{m-1} 2^{-1} + j_m 2^{-2}} |1\rangle) \otimes \frac{1}{\sqrt{2}} (|0\rangle + \omega^{j_m 2^{-1}} |1\rangle).$$

Set

$$R_k = \begin{pmatrix} 1 & 0 \\ 0 & \omega^{2^k} \end{pmatrix}, \tag{2.10}$$

then (2.9) is obtained as follows: first apply H to $(\mathbb{C}^2)_1$ then a linear map $\Lambda^1 R_j$, defined by $|x\rangle \otimes |y\rangle \mapsto |x\rangle \otimes R_j |y\rangle$ if $|x\rangle \neq |0\rangle$ and to $|x\rangle \otimes |y\rangle$ if $|x\rangle = |0\rangle$, to $(\mathbb{C}^2)_j \otimes (\mathbb{C}^2)_1$ for $j = 2, \ldots, m$. Note that at this point only the $(\mathbb{C}^2)_1$-term has been altered. From now on leave the $(\mathbb{C}^2)_1$-slot alone. Next apply H to $(\mathbb{C}^2)_2$ then $\Lambda^1 R_{j-1}$ to $(\mathbb{C}^2)_j \otimes (\mathbb{C}^2)_2$ for $j = 3, \ldots, m$. Then apply H to $(\mathbb{C}^2)_3$ then $\Lambda^1 R_{j-2}$ to $(\mathbb{C}^2)_j \otimes (\mathbb{C}^2)_3$ for $j = 4, \ldots, m$. Continue, until finally one just applies H to $(\mathbb{C}^2)_m$. Finally to obtain the DFT, reverse the orders of the factors (a classical operation).

In practice, one has to fix a quantum gate set, i.e., a finite set of unitary operators that will be allowed in algorithms, in advance. Thus in general it will be necessary to approximate the transformations R_k from elements of our gate set, so one only obtains an approximation of the DFT.

2.4.3 The Hidden Subgroup Problem

Given a discrete group G with a specific representation of its elements in binary, a function $f : G \to \mathbb{F}_2^n$, and a device that computes f (for unit cost), and the knowledge that there exists a subgroup $G' \subset G$ such that $f(x) = f(y)$ if and only if $xy^{-1} \in G'$, find G'.

For finitely generated abelian groups, it is sufficient to solve the problem for $G = \mathbb{Z}^{\oplus k}$ as all finitely generated abelian groups are quotients of some $\mathbb{Z}^{\oplus k}$.

Simons algorithm is for the hidden subgroup problem with $G = \mathbb{Z}_2^{\oplus m}$, see [15, §13.1]. The DFT_2 matrix is just

$$H = \frac{1}{\sqrt{2}} \begin{pmatrix} 1 & -1 \\ -1 & 1 \end{pmatrix}$$

and G' is the subgroup generated by $a \in \mathbb{Z}_2^{\oplus m}$.

Shor's algorithm for factoring (after classical preparation) amounts to the case $G = \mathbb{Z}$ and F is the function $x \mapsto a^x \bmod N$. It has generated intense interest in quantum computation because no classical or probabilistic polynomial time algorithm for factoring is known. For example, most "secure" electronic communication is based on the difficulty of factoring a number into its prime factors, so the real world impact of a quantum computer would be substantial. See, e.g., http://www.math.tamu.edu/~jml/CNSA-Suite-and-Quantum-Computing-FAQ.pdf. See, e.g., [2, §10.6] for an exposition of Shor's algorithm.

2.5 Classical Information Theory

Quantum information theory is based on classical information theory, so I review the classical theory. The discovery/invention of the bit by Tukey and its development by Shannon [22] was one of the great scientific achievements of the twentieth century, as it changed the way one views information, giving it an abstract formalism that is discussed in this section. The link to quantum information is explained in Sect. 2.7.

The basic question is: given a physical channel, e.g., a telegraph wire, what is the maximum rate of transmission of messages allowing for a small amount of error? I begin with toy examples, leading up to Shannon's two fundamental theorems.

2.5.1 Data Compression: Noiseless Channels

(Following [6]) A source emits symbols x from an alphabet $\mathcal{X}$ that we want to store efficiently so we try to encode x in a small number of bits, to say $y \in \mathcal{Y}$ in a way that one can decode it later to recover x (Fig. 2.2).

The symbols from $\mathcal{X}$ do not necessarily occur with the same frequency. Let $p = P_{\mathcal{X}}$ denote the associated probability distribution. What is the minimum possible size for $\mathcal{Y}$? Since we are dealing in bits, it will be convenient to use the logarithms of cardinalities, so define the *capacity* as $\mathrm{Cap}(P_{\mathcal{X}}) := \min \log |\mathcal{Y}|$.

Fig. 2.2 Message from source encoded into bits then decoded

Consider the case $\mathcal{X} = \{a, b, c, d\}$ where $p(a) = 0.1$, $p(b) = 0$, $p(c) = 0.4$ and $p(d) = 0.5$. One can clearly get away with $|\mathcal{Y}| = 3$, e.g., for the encoder, send a, b to 1, c to 2 and d to 3, then for the decoder, send 1 to a, 2 to c and 3 to d. In general, one can always throw away symbols with probability zero. On the other hand, one cannot map two distinct symbols that do occur to the same symbol, as there would be no way to distinguish them when decoding. Thus $\mathrm{Cap}(p) = \log \mathrm{supp}(p)$, where $\mathrm{supp}(p) = \#\{x \in \mathcal{X} \mid p(x) > 0\}$.

Now say we are willing to tolerate a small error. First rephrase what we did probabilistically: Let $p^{enc}(y|x)$ denote the conditional probability distribution of the encoder $\mathcal{E}$ and $p^{dec}(x|y)$ that of the decoder $\mathcal{D}$. Our requirement was for all x,

$$p[x = \mathcal{D} \circ \mathcal{E}(x)] = \sum_{y,x'} p^{enc}(y|x)p^{dec}(x'|y)\delta_{x,x'} = 1.$$

Now relax it to

$$\sum_{x,y,x'} p(x)p^{enc}(y|x)p^{dec}(x'|y)\delta_{x,x'} \geq 1 - \epsilon.$$

for some error ϵ that we are willing to tolerate. In addition to throwing out the symbols that do not appear, we may also discard the largest set of symbols whose total probability is smaller than ϵ. Call the corresponding quantity $\mathrm{Cap}^{\epsilon}(p)$. In this example, if one takes $\epsilon > 0.1$, one can lower storage cost, taking $|\mathcal{Y}| = 2$.

Recall that a probability distribution $p : \mathcal{X} \to [0, 1]$ must satisfy $\sum_{x \in \mathcal{X}} p(x) = 1$. Relax this to *non-normalized* probability distributions, $q : \mathcal{X} \to [0, 1]$, where $\sum_{x \in \mathcal{X}} q(x) \leq 1$. We obtain: $\mathrm{Cap}^{\epsilon}(p) = \min \log \mathrm{supp}(q)$, where the min is taken over all non-normalized probability distributions q satisfying $q(x) \leq p(x)$ and $\sum_{x \in \mathcal{X}} q(x) \geq 1 - \epsilon$.

Now say we get not a single symbol, but a string of n symbols, so we seek an encoder $\mathcal{E} : \mathcal{X}^n \to \mathcal{Y}(n)$, where $\mathcal{Y}(n)$ is a set that varies with n, and decoder $\mathcal{D} : \mathcal{Y}(n) \to \mathcal{X}^n$, and we want to minimize $|\mathcal{Y}(n)|$, with a tolerance of error that goes to zero as n goes to infinity. In practice one wants to send information through a communication channel (e.g. telegraph wire). The channel can only send a limited number of bits per second, and we want to maximize the amount of information we can send per second: $\lim_{\epsilon \to 0} \lim_{n \to \infty} \frac{1}{n} \mathrm{Cap}^{\epsilon}(p^n)$.

The string $x_1 \cdots x_n =: \overline{x}^n$ is identically and independently distributed (i.i.d), that is each x_j is drawn from the same probability distribution and the draw of x_j is independent of the draws of the other x_i. Say $\mathcal{X} = \{1, \ldots, d\}$ with $p(j) = p_j$. The probability of any given string occurring depends only on the number of 1's 2's etc. in the string and not on their order. A string with c_j j's occurs with probability $p_1^{c_1} \cdots p_d^{c_d}$. (Note that $c_1 + \cdots + c_d = n$.) The number of strings with this probability is

$$\binom{n}{c_1, \ldots, c_d} := \frac{n!}{c_1! \cdots c_d!}$$

and we need to estimate this quantity.

Stirling's formula implies $\ln(n!) = n \ln(n) - n + O(\ln(n))$. In particular, for $0 < \beta < 1$ such that $\beta n \in \mathbb{Z}$,

$$\log \binom{n}{\beta n} = n[-\beta \log(\beta) - (1 - \beta) \log(1 - \beta)] + O(\log(n)).$$

Let $H(\beta) = -\beta \log(\beta) - (1 - \beta) \log(1 - \beta)$ and more generally, for $\overline{p} = (p_1, \ldots, p_d)$, let

$$H(\overline{p}) = - \sum_{i=1}^{d} p_i \log(p_i),$$

the *Shannon entropy* of $\overline{p}$. It plays a central role in information theory.

Define a map $wt : \mathcal{X}^n \to \mathbb{R}^d$ by $\overline{x}^n \mapsto (c_1, \ldots, c_d)$, where c_j is the number of j's appearing in $\overline{x}^n$. Then the expectation is $E[wt(\overline{x}^n)] = (np_1, \ldots, np_d)$. The weak law of large numbers states that for any $\epsilon > 0$,

$$\lim_{n \to \infty} p[||\frac{1}{n} wt(\overline{x}^n) - E[wt(\overline{x}^n))]||_1 > \epsilon] = 0$$

where for $f : \mathcal{Z} \to \mathbb{R}^d$, define $||f||_1 = \sum_{z \in \mathcal{Z}} |f(z)|$. In our case, $\mathcal{Z} = \mathcal{X}^n$.

Now simply throw out all strings $\overline{x}^n$ with $||\frac{1}{n}(wt(\overline{x}^n) - E[wt(\overline{x}^n))]||_1 > \epsilon$, and take $\mathcal{Y}(n)$ of size

$$|\mathcal{Y}(n)| = \#\{\overline{x}^n \mid ||\frac{1}{n}(wt(\overline{x}^n) - E[wt(\overline{x}^n))]||_1 < \epsilon\}$$

$$= \sum_{\substack{\overline{x}^n| \\ ||\frac{1}{n}(wt(\overline{x}^n) - E[wt(\overline{x}^n))]||_1 < \epsilon}} \binom{n}{wt(\overline{x}^n)}.$$

If ϵ is small, the multinomial coefficients appearing will all be very close to

$$\binom{n}{np_1, \ldots, np_d}$$

and for what follows, one can take the crude approximation

$$|\mathcal{Y}(n)| \leq poly(n)\binom{n}{np_1, \ldots, np_d} \tag{2.11}$$

(recall that d is fixed).

Taking logarithms, the right hand side of (2.11) becomes $nH(\overline{p}) + O(\log(n))$. Thus

$$\frac{1}{n} \log |\mathcal{Y}(n)| \leq H(\overline{p}) + o(1)$$

and $\lim_{\epsilon \to 0} \lim_{n \to \infty} \frac{1}{n} \text{Cap}^\epsilon(p^n) \leq H(\overline{p})$.

Theorem 2.5.1 ([22]) $\lim_{\epsilon \to 0} \lim_{n \to \infty} \frac{1}{n} \text{Cap}^\epsilon(p^n) = H(\overline{p})$.

The full proof uses the law of large numbers.

2.5.2 Transmission over Noisy Channels

Say symbols x are transmitted over a channel subject to noise, and symbols y are received so one may or may not have $y = x$. Intuitively, if the noise is small, with some redundancy it should be possible to communicate accurate messages most of the time. In a noiseless channel the maximal rate of transmission is just $H(p_\mathcal{X})$, but now we must subtract off something to account for the uncertainty that, upon receiving y, that it was the signal sent. This something will be the *conditional entropy*: Recall the conditional probability of i occurring given knowledge that j occurs (assuming $p(j) > 0$): $p_{\mathcal{X}|\mathcal{Y}}(i|j) = \frac{p_{\mathcal{X},\mathcal{Y}}(i,j)}{p_\mathcal{Y}(j)}$ (also recall $p_\mathcal{Y}(j) = \sum_i p_{\mathcal{X},\mathcal{Y}}(i,j)$). Define the conditional entropy

$$H(\overline{p}_\mathcal{Y}|\overline{p}_\mathcal{X}) := -\sum_{i,j} p_{\mathcal{X},\mathcal{Y}}(i,j) \log p_{\mathcal{Y}|\mathcal{X}}(j|i).$$

Note that

$$H(\overline{p}_\mathcal{Y}|\overline{p}_\mathcal{X}) = H(\overline{p}_{\mathcal{X},\mathcal{Y}}) - H(\overline{p}_\mathcal{X}) \tag{2.12}$$

or equivalently $H(\overline{p}_{\mathcal{X},\mathcal{Y}}) = H(\overline{p}_\mathcal{X}) + H(\overline{p}_\mathcal{Y}|\overline{p}_\mathcal{X})$, the uncertainty of $p_{\mathcal{X},\mathcal{Y}}$ is the uncertainty of $p_\mathcal{X}$ plus the uncertainty of $p_\mathcal{Y}$ given $p_\mathcal{X}$. In particular $H(\overline{p}_\mathcal{Y}) \geq H(\overline{p}_\mathcal{Y}|\overline{p}_\mathcal{X})$, i.e., with extra knowledge, our uncertainty about $p_\mathcal{Y}$ cannot increase, and decreases unless $p_\mathcal{X}$ and $p_\mathcal{Y}$ are independent.

2.5.2.1 Capacity of a Noisy Channel

Define the *capacity* of a noisy channel to be the maximum rate over all possible probability distributions on the source:

$$\text{Cap} := \max_{q_X} \left(H(q_X) - H(q_X | p_Y) \right).$$

Shannon [22] proves that Cap lives up to its name: if the entropy of a discrete channel is below Cap then there exists an encoding $\overline{p}$ of the source such that information can be transmitted over the channel with an arbitrarily small frequency of errors. The basic idea is the same as the noiseless case, however there is a novel feature that now occurs frequently in complexity theory arguments—that instead of producing an algorithm to find the efficient encoding, Shannon showed that a *random* choice of encoding will work.

After presenting the proof, Shannon remarks: "An attempt to obtain a good approximation to ideal coding by following the method of the proof is generally impractical. ... Probably this is no accident but is related to the difficulty of giving an explicit construction for a good approximation to a random sequence". To my knowledge, this is the first time that the difficulty of "finding hay in a haystack" (phrase due to Howard Karloff) is mentioned in print. This problem is central to complexity: for example, Valiant's algebraic version of $\mathbf{P} \neq \mathbf{NP}$ can be phrased as the problem of finding a sequence of explicit polynomials that are difficult to compute, while it is known that a random sequence is indeed difficult to compute. According to A. Wigderson, the difficulty of writing down random objects was also explicitly discussed by Erdös, in the context of random graphs, at least as early as 1947, in relation to his seminar paper [11]. This paper, along with [22] gave rise to the now ubiquitous probabilistic method in complexity theory.

2.6 Reformulation of Quantum Mechanics

I discuss two inconveniences about our formulation of the postulates of quantum mechanics, leading to a reformulation of the postulates in terms of density operators.

2.6.1 Partial Measurements

A measurement of a state $|\psi\rangle = \sum z_I |I\rangle$ was defined as a procedure that gives us $I = (i_1, \ldots, i_n) \in \{0, 1\}^n$ with probability $|z_I|^2$. But in our algorithms, this is not what happened: we were working not in $(\mathbb{C}^2)^{\otimes n}$, but $(\mathbb{C}^2)^{\otimes n+m}$ where there were m "workspace" qubits we were not interested in measuring. So our measurement was more like the projections onto the *spaces* $|I\rangle \otimes (\mathbb{C}^2)^{\otimes m}$. I now define this generalized notion of measurement.

To make the transition, first observe that $|z_I|^2 = \langle\psi|\Pi_I|\psi\rangle$, where Π_I : $(\mathbb{C}^2)^{\otimes n} \to \mathbb{C}|I\rangle$ is the orthogonal projection onto the line spanned by $|I\rangle$.

Say we are only interested in the first n bits of a system of $n + m$ bits, and want to know the probability a measurement gives rise to some I represented by a vector $|I\rangle \in (\mathbb{C}^2)^{\otimes n}$, but we have $|\psi\rangle \in (\mathbb{C}^2)^{\otimes n+m}$. Adopt the notation $|\phi\rangle\langle\psi| := |\phi\rangle\otimes\langle\psi|$. Then the probability of obtaining $|I\rangle$ given $|\psi\rangle$ is

$$
\begin{aligned}
p(|I\rangle \mid |\psi\rangle) &= \sum_{J \in \{0,1\}^m} p(|\psi\rangle, |IJ\rangle) \\
&= \sum_J \langle\psi|IJ\rangle\langle IJ|\psi\rangle \\
&= \langle\psi|(|I\rangle\langle I|\otimes \mathrm{Id}_{(\mathbb{C}^2)^{\otimes m}})|\psi\rangle \\
&= \langle\psi|\Pi_{\mathcal{M}}|\psi\rangle
\end{aligned}
$$

where $\Pi_{\mathcal{M}} : (\mathbb{C}^2)^{\otimes n+m} \to |I\rangle\otimes(\mathbb{C}^2)^{\otimes m} =: \mathcal{M}$ is the orthogonal projection operator. With this definition, one can allow $\mathcal{M} \subset \mathcal{H}$ to be *any* linear subspace, which will simplify our measurements. (Earlier, if we wanted to measure the probability of a non-basis state, we had to change bases before measuring.) Write $p_\psi(\mathcal{M}) := \langle\psi|\Pi_{\mathcal{M}}|\psi\rangle$ for the probability of measuring $|\psi\rangle$ in state $\mathcal{M}$.

One may think of projection operators as representing outside interference of a quantum system, like adding a filter to beams being sent that destroy states not in $\mathcal{M}$. Recall that in classical probability, one has the identity:

$$p(M_1 \cup M_2) = p(M_1) + p(M_2) - p(M_1 \cap M_2). \tag{2.13}$$

The quantum analog is *false* in general: Let $\mathcal{H} = \mathbb{C}^2$, $\mathcal{M}_1 = \mathbb{C}|0\rangle$ and $\mathcal{M}_2 = \mathbb{C}(|0\rangle + |1\rangle)$ Let $|\psi\rangle = \alpha|0\rangle + \beta|1\rangle$ with $|\alpha|^2 + |\beta|^2 = 1$. Then (and in general) $p_\psi(\mathrm{span}\{\mathcal{M}_1, \mathcal{M}_2\}) \neq p_\psi(\mathcal{M}_1) + p_\psi(\mathcal{M}_2) - p_\psi(\mathcal{M}_1 \cap \mathcal{M}_2)$.

However, one can recover (2.13) if the projection operators commute:

Proposition 2.6.1 *If* $\Pi_{\mathcal{M}_1}\Pi_{\mathcal{M}_2} = \Pi_{\mathcal{M}_2}\Pi_{\mathcal{M}_1}$ *then for all* ψ, $p_\psi(\mathrm{span}\{\mathcal{M}_1, \mathcal{M}_2\}) = p_\psi(\mathcal{M}_1) + p_\psi(\mathcal{M}_2) - p_\psi(\mathcal{M}_1 \cap \mathcal{M}_2)$.

2.6.2 Mixing Classical and Quantum Probability

A typical situation in probability is as follows: you want a cookie, but can't make up your mind which kind, so you decide to take one at random from the cookie jar to eat. However when you open the cupboard, you find there are two different cookie jars H and T, each with a different distribution of cookies, say P_H and P_T. You decide to flip a coin to decide which jar and say your coin is biased with probability

p for heads (choice H). The resulting probability distribution is

$$pP_H + (1 - p)P_T.$$

Let's encode this scenario with vectors. Classically, if vectors corresponding to P_H, P_T are respectively v_H, v_T, the new vector is $pv_H + (1 - p)v_T$. The probability of drawing a chocolate chip (CC) cookie is $pP_H(CC) + (1 - p)P_T(CC) = pv_{H,CC} + (1 - p)v_{T,CC}$.

But what should one take in generalized probability (where one uses the ℓ_2 norm instead of the ℓ_1 norm)? Given $|\psi_A\rangle = \sum z_I|I\rangle$, $|\psi_B\rangle = \sum w_J|J\rangle$, we want to make a measurement that gives us $p|z_{CC}|^2 + (1-p)|w_{CC}|^2$. Unfortunately $|pz_{CC} + (1 - p)w_{CC}|^2 \neq p|z_{CC}|^2 + (1 - p)|w_{CC}|^2$ in general. To fix this problem one enlarges the notion of state and further modifies the definition of measurement.

Our problem comes from having a mixture of ℓ_1 and ℓ_2 norms. The fix will be to rewrite $|\psi\rangle$ in a way that the ℓ_2 norm becomes an ℓ_1 norm. That is, construct an object that naturally contains the squares of the norms of the coefficients of $|\psi_A\rangle$. Consider the endomorphism $|\psi_A\rangle\langle\psi_A| = \sum_{I,J} z_I \bar{z}_J|I\rangle\langle J|$. It is rank one, and in the standard basis its diagonal entries are the quantities we want.

To measure them, let Π_J denote the projection onto the J-th coordinate. Then

$$\text{trace}(\Pi_J|\psi_A\rangle\langle\psi_A|) = |z_{A,J}|^2$$

is the desired quantity.

Now back to our cookie jars, set

$$\rho = p|\psi_A\rangle\langle\psi_A| + (1 - p)|\psi_B\rangle\langle\psi_B|$$

and observe that

$$\text{trace}(\Pi_J\rho) = p|z_{A,J}|^2 + (1 - p)|z_{B,J}|^2$$

as desired.

Given a finite set of states $\{|\psi_1\rangle, \ldots, |\psi_s\rangle\}$, with $p(|\psi_i\rangle) = p_i$, and $\sum_i p_i = 1$, set $\rho = \sum_k p_k|\psi_k\rangle\langle\psi_k| \in \text{End}(\mathcal{H})$. Note that ρ has the properties

(1) $\rho = \rho^\dagger$, i.e., ρ is Hermitian,
(2) $\forall|\eta\rangle$, $\langle\eta|\rho|\eta\rangle \geq 0$, i.e., ρ is *positive*,
(3) $\text{trace}(\rho) = 1$.

This motivates the following definition:

Definition 2.6.2 An operator $\rho \in \text{End}(\mathcal{H})$ satisfying 1,2,3 above is called a *density operator*.

Note that a density operator that is diagonal in the standard basis of $\mathbb{C}^d$ corresponds to a probability distribution on $\{1, \ldots, d\}$, so the definition includes classical probability as well as our old notion of state (which are the rank one density operators). The set of density operators is invariant under the induced action of $U(\mathcal{H})$ on $\mathrm{End}(\mathcal{H})$.

Different scenarios can lead to the same density operator. However, two states with the same density operator are physically indistinguishable.

2.6.3 Reformulation of the Postulates of Quantum Mechanics

Postulate 1 Associated to any isolated physical system is a Hilbert space $\mathcal{H}$, call the *state space*. The system is described by its density operator $\rho \in \mathrm{End}(\mathcal{H})$.

Postulate 2 The evolution of an isolated system is described by the action of unitary operators on ρ.

Postulate 3 Measurements correspond to a collection of projection operators $\Pi_{\mathcal{M}_j}$ such that $\sum_k \Pi_{\mathcal{M}_k} = \mathrm{Id}_{\mathcal{H}}$. The probability that ρ is in measured in state $\mathcal{M}_j$ is $\mathrm{trace}(\Pi_{\mathcal{M}_j}\rho)$. Such measurements are called "Positive Operator-Valued Measurements", or POVM, in the literature.

Sometimes it is convenient to allow more general measurements than POVM:

Postulate 3′ Projective measurements correspond to a collection of Hermitian operators $X_j \in \mathrm{End}\,\mathcal{H}$ such that $\sum_k X_k = \mathrm{Id}_{\mathcal{H}}$. The probability that ρ is in measured in state X_j is $\mathrm{trace}(X_j\rho)$.

Postulate 4 regarding composite systems is unchanged.

Remark 2.6.3 Note that for $A \in \mathrm{End}\,\mathcal{H} = \mathcal{H}^*{\otimes}\mathcal{H}$, $\mathrm{trace}(A)$ is the image of A under the contraction map $\mathcal{H}^*{\otimes}\mathcal{H} \to \mathbb{C}$, $\langle v| \otimes |w\rangle \mapsto \langle v|w\rangle$. For $A \in \mathrm{End}(\mathcal{H}_1{\otimes}\mathcal{H}_2) = (\mathcal{H}_1^*{\otimes}\mathcal{H}_2^*){\otimes}(\mathcal{H}_1{\otimes}\mathcal{H}_2)$, define the partial trace $\mathrm{trace}_{\mathcal{H}_1}(A)$ to be the image of A under the contraction $\mathcal{H}_1^*{\otimes}\mathcal{H}_2^*{\otimes}\mathcal{H}_1{\otimes}\mathcal{H}_2 \to \mathcal{H}_2^*{\otimes}\mathcal{H}_2$ given by $\langle\phi|\otimes\langle\psi|\otimes|v\rangle\otimes|w\rangle \mapsto \langle\phi|v\rangle\langle\psi|\otimes|w\rangle = \langle\phi|v\rangle|w\rangle\langle\psi|$.

2.6.4 Expectation and the Uncertainty Principle

Let $A \in \mathrm{End}(\mathcal{H})$ be a Hermitian operator with eigenvalues $\lambda_1, \ldots, \lambda_k$ and eigenspaces $\mathcal{M}_j$. If our system is in state ρ, one can consider A as a random variable that takes the value λ_j with probability $\mathrm{trace}(\Pi_{\mathcal{M}_j}\rho)$.

The expectation of a random variable $X : \mathcal{X} \to \mathbb{R}$ is $E[X] := \sum_{j\in\mathcal{X}} X(j)p(j)$.

If a system is in state ρ, the expectation of a Hermitian operator $A \in \mathrm{End}(\mathcal{H})$ is $\mathrm{trace}(A\rho)$ because $E[A] = \sum_{\lambda_j} \lambda_j \, \mathrm{trace}(\Pi_{\mathcal{M}_j}\rho) = \mathrm{trace}((\sum_{\lambda_j} \lambda_j \Pi_{\mathcal{M}_j})\rho) = \mathrm{trace}(A\rho)$.

One way mathematicians describe the famous Heisenberg uncertainty principle is that it is impossible to localize both a function and its Fourier transform. Another interpretation comes from probability:

First note that given a random variable, or Hermitian operator X, one can replace it with an operator of mean zero $\hat{X} := X - E(X\rho)$ Id. For notational convenience, I state the uncertainty principle for such shifted operators.

The variance $var(X)$ of a random variable is $var(X) = E[X - E(X)]^2$. The standard deviation $\sigma(X) = \sqrt{var(X)}$ of X is a measure of the failure of the corresponding probability distribution to be concentrated at a point, i.e., failure of the induced probability distribution to have a certain outcome.

Proposition 2.6.4 *Let X, Y be Hermitian operators of mean zero, corresponding to observables on a system in state ρ, let Then*

$$\sigma(X)\sigma(Y) \geq \frac{|\,\mathrm{trace}([X, Y]\rho)|}{2}.$$

The uncertainty principle says that the failure of two Hermitian operators to commute lower bounds the product of their uncertainties. In particular, if they do not commute, neither can give rise to a classical (certain) measurement. It is a consequence of the Cauchy-Schwarz inequality.

2.6.5 Pure and Mixed States

Definition 2.6.5 Let $\rho \in \mathrm{End}(\mathcal{H})$ be a density operator. If $\mathrm{rank}(\rho) = 1$, i.e. $\rho = |\xi\rangle\langle\xi|$, ρ is called a *pure state*, and otherwise it is called a *mixed state*.

The partial trace of a pure state can be a mixed state. For example, if $\rho = |\psi\rangle\langle\psi|$ with $\psi = \frac{1}{\sqrt{2}}(|00\rangle + |11\rangle) \in \mathcal{H}_1\otimes\mathcal{H}_2$, then $\mathrm{trace}_{\mathcal{H}_2}(\rho) = \frac{1}{2}(|0\rangle\langle0| + |1\rangle\langle1|)$.

The following proposition shows that one could avoid density operators altogether by working on a larger space:

Proposition 2.6.6 *An arbitrary mixed state $\rho \in \mathrm{End}(\mathcal{H})$ can be represented as the partial trace $\mathrm{trace}_{\mathcal{H}'}|\psi\rangle\langle\psi|$ of a pure state in $\mathrm{End}(\mathcal{H}\otimes\mathcal{H}')$ for some Hilbert space $\mathcal{H}'$. In fact, one can always take $\mathcal{H}' = \mathcal{H}^*$.*

Given a density operator $\rho \in \mathrm{End}(\mathcal{H})$, there is a well defined operator $\sqrt{\rho} \in \mathrm{End}(\mathcal{H})$ whose eigenvectors are the same as for ρ, and whose eigenvalues are the positive square roots of the eigenvalues of ρ. To prove the proposition, given $\rho \in \mathcal{H}\otimes\mathcal{H}^*$, consider $|\sqrt{\rho}\rangle\langle\sqrt{\rho}| \in \mathrm{End}(\mathcal{H}\otimes\mathcal{H}^*)$. Then $\rho = \mathrm{trace}_{\mathcal{H}^*}(|\sqrt{\rho}\rangle\langle\sqrt{\rho}|)$. A pure state whose partial trace is ρ is called a *purification* of ρ.

2.7 Communication Across a Quantum Channel

Now instead of having a source $\mathcal{X}^{\times n}$ our "source" is $\mathcal{H}^{\otimes n}$, where one can think of $\mathcal{H}^{\otimes n} = \mathcal{H}_A^{\otimes n}$, and Alice will "transmit" a state to Bob, and instead of a probability distribution p one has a density operator ρ.

What is a quantum channel? It should be a linear map sending $\rho \in \mathrm{End}(\mathcal{H}_A)$ to some $\Phi(\rho) \in \mathrm{End}(\mathcal{H}_B)$.

First consider the special case $\mathcal{H}_A = \mathcal{H}_B$. One should allow coupling with an auxiliary system, i.e.,

$$\rho \mapsto \rho \otimes \sigma \in \mathrm{End}(\mathcal{H}_A \otimes \mathcal{H}_C). \tag{2.14}$$

One should also allow the state $\rho \otimes \sigma$ to evolve in $\mathrm{End}(\mathcal{H}_A \otimes \mathcal{H}_C)$, i.e., be acted upon by an arbitrary $U \in \mathbf{U}(\mathcal{H}_A \otimes \mathcal{H}_C)$. Finally one should allow measurements, i.e., tracing out the $\mathcal{H}_C$ part. In summary, a quantum channel $\mathcal{H}_A \to \mathcal{H}_A$ is a map of the form $\rho \mapsto \mathrm{trace}_{\mathcal{H}_C}(U(\rho \otimes \sigma)U^{-1})$. More generally to go from $\mathcal{H}_A$ to $\mathcal{H}_B$, one needs to allow isometries as well. Such maps are the *completely positive trace preserving maps* (CPTP), where a map Λ is *completely positive* if $\Lambda \otimes \mathrm{Id}_{\mathcal{H}_E}$ is positive for all $\mathcal{H}_E$.

We seek an encoder $\mathcal{E}$ and decoder $\mathcal{D}$ and a compression space $\mathcal{H}_{0n}$:

$$\mathcal{H}^{\otimes n} \overset{\mathcal{E}}{\to} \mathcal{H}_{0n} = (\mathbb{C}^2)^{\otimes nR} \overset{\mathcal{D}}{\to} \mathcal{H}^{\otimes n}$$

with R as small as possible such that $\mathcal{E} \circ \mathcal{D}(\rho^{\otimes n})$ converges to $\rho^{\otimes n}$ as $n \to \infty$. To determine R, we need a quantum version of entropy.

Definition 2.7.1 The *von Neumann entropy* of a density operator ρ is $H(\rho) = -\mathrm{trace}(\rho \log(\rho))$.

Here $\log(\rho)$ is defined as follows: write ρ in terms of its eigenvectors and eigenvalues, $\rho = \sum_j \lambda_j |\psi_j\rangle\langle\psi_j|$, then $\log(\rho) = \sum_j \log(\lambda_j)|\psi_j\rangle\langle\psi_j|$.

If $\rho = \sum_j \lambda_j |\psi_j\rangle\langle\psi_j|$, then $H(\rho) = -\sum_j \lambda_j \log(\lambda_j)$ so if ρ is classical (i.e., diagonal), one obtains the Shannon entropy.

Proposition 2.7.2 *The von Neumann entropy has the following properties:*

(1) $H(\rho) \geq 0$ *with equality if and only if ρ is pure.*
(2) *Let $\dim \mathcal{H} = d$. Then $H(\rho) \leq \log(d)$ with equality if and only if $\rho = \frac{1}{d}\mathrm{Id}_{\mathcal{H}}$.*
(3) *If $\rho = |\psi\rangle\langle\psi| \in \mathrm{End}(\mathcal{H}_A \otimes \mathcal{H}_B)$, then $H(\rho_A) = H(\rho_B)$, where $\rho_A = \mathrm{trace}_{\mathcal{H}_B}(\rho) \in \mathrm{End}(\mathcal{H}_A)$.*

Notice that in particular von Neumann entropy is maximized for $|epr\rangle$. In Sects. 2.8 and 2.9 I discuss entanglement as a resource and von Neumann entropy as a measurement of that resource.

Theorem 2.7.3 ([21], The Quantum Noiseless Channel Theorem) *Let $(\mathcal{H}, \rho)$ be an i.i.d. quantum source. If $R > H(\rho)$, then there exists a reliable compression* scheme of rate R. That is, there exists a compression space $\mathcal{H}_{0n}$, of dimension 2^{nR}, and encoder $\mathcal{E} : \mathcal{H}^{\otimes n} \to \mathcal{H}_{0n}$ and a decoder $\mathcal{D} : \mathcal{H}_{0n} \to \mathcal{H}^{\otimes n}$ such that $\mathcal{D} \circ \mathcal{E}(\rho^{\otimes n})$ converges to $\rho^{\otimes n}$ as $n \to \infty$. If $R < H(\rho)$, then any compression scheme is unreliable.

2.8 More on von Neumann Entropy and Its Variants

First for the classical case, define the relative entropy $H(\overline{p}\|\overline{q}) := -\sum p_i \log \frac{q_i}{p_i} = -H(\overline{p}) - \sum_i p_i \log(q_i)$. It is zero when $\overline{p} = \overline{q}$ and is otherwise positive. Define the relative von Neumann entropy $H(\rho\|\sigma) := \operatorname{trace}(\rho \log(\rho)) - \operatorname{trace}(\rho \log(\sigma))$. It shares the positivity property of its classical cousin: [16] $H(\rho\|\sigma) \geq 0$ with equality if and only if $\rho = \sigma$.

Proposition 2.8.1 (von Neumann Entropy Is Non-decreasing Under Projective Measurements) *Let Π_i be a complete set of orthogonal projectors, set $\rho' = \sum_i \Pi_i \rho \Pi_i$. Then $H(\rho') \geq H(\rho)$ with equality if and only if $\rho' = \rho$.*

If we think of the entropy of ρ as a measurement of entanglement, i.e., a measurement of ρ as a communication resource, we see this resource decreases after a measurement.

Proof First note that $0 \leq H(\rho\|\rho') = -H(\rho) - \operatorname{trace}(\rho \log(\rho'))$. Now

$$\operatorname{trace}(\rho \log(\rho')) = \operatorname{trace}\left(\sum_i \Pi_i \rho \log(\rho')\right)$$

$$= \operatorname{trace}\left(\sum_i \Pi_i \rho \log(\rho')\Pi_i\right)$$

because $\Pi_i^2 = \Pi_i$ and $\operatorname{trace}(AB) = \operatorname{trace}(BA)$. Now Π_i commutes with ρ' and $\log(\rho')$ because $\Pi_i \Pi_j = 0$ if $i \neq j$, so

$$\operatorname{trace}(\rho \log(\rho')) = \operatorname{trace}(\sum_i \Pi_i \rho \Pi_i \log(\rho'))$$

$$= \operatorname{trace}(\rho' \log(\rho'))$$

$$= -H(\rho')$$

Putting it all together, we obtain the result. ◻

Here and in what follows ρ_{AB} is a density operator on $\mathcal{H}_A \otimes \mathcal{H}_B$ and $\rho_A = \operatorname{trace}_{\mathcal{H}_B}(\rho_{AB})$, $\rho_B = \operatorname{trace}_{\mathcal{H}_A}(\rho_{AB})$ are respectively the induced density operators on $\mathcal{H}_A$, $\mathcal{H}_B$.

von Neumann entropy is sub-additive: $H(\rho_{AB}) \leq H(\rho_A) + H(\rho_B)$ with equality if and only if $\rho_{AB} = \rho_A \otimes \rho_B$. It also satisfies a triangle inequality: $H(\rho_{AB}) \geq |H(\rho_A) - H(\rho_B)|$.

Recall the conditional Shannon entropy is defined to be $H(\overline{p}_{\mathcal{X}} | \overline{p}_{\mathcal{Y}}) = -\sum_{i,j} p_{\mathcal{X} \times \mathcal{Y}}(i, j) \log p_{\mathcal{X}|\mathcal{Y}}(i|j)$, the entropy of $p_{\mathcal{X}}$ conditioned on $y = j$, averaged over $\mathcal{Y}$. It is not clear how to "condition" one density matrix on another, so one needs a different definition. Recall that Shannon entropy satisfies $H(\overline{p}_{\mathcal{X}} | \overline{p}_{\mathcal{Y}}) = H(\overline{p}_{\mathcal{X} \times \mathcal{Y}}) - H(\overline{p}_{\mathcal{Y}})$, and the right hand side of this expression does make sense for density operators, so define, for ρ_{AB} a density operator on $\mathcal{H}_A \otimes \mathcal{H}_B$,

$$H(\rho_A | \rho_B) := H(\rho_{AB}) - H(\rho_B). \tag{2.15}$$

Note that $H(\rho_A | \rho_B)$ is a function of ρ_{AB}, as $\rho_B = \text{trace}_{\mathcal{H}_A} \rho_{AB}$.

WARNING: it is possible that the conditional von Neumann entropy is *negative* as it is possible that $H(\rho_B) > H(\rho_{AB})$. Consider the following example: Let $|\psi\rangle = \frac{1}{\sqrt{2}}(|00\rangle + |11\rangle) \in \mathcal{H}_A \otimes \mathcal{H}_B$. Then $\rho_A = \frac{1}{2} \text{Id}_{\mathcal{H}_A} = \frac{1}{2}(|0\rangle\langle0| + |1\rangle\langle1|)$ so $H(\rho_A) = 1$, but $H(|\psi\rangle\langle\psi|) = 0$ because $|\psi\rangle\langle\psi|$ is pure.

However, vestiges of positivity are true in the quantum case:

Theorem 2.8.2 (Strong Sub-additivity) *Let ρ_{ABC} be a density operator on $\mathcal{H}_A \otimes \mathcal{H}_B \otimes \mathcal{H}_C$. Then*

$$H(\rho_C | \rho_A) + H(\rho_C | \rho_B) \geq 0 \tag{2.16}$$

and

$$H(\rho_{ABC}) - [H(\rho_{AB}) + H(\rho_{BC})] + H(\rho_B) \geq 0. \tag{2.17}$$

Strong sub-additivity has many consequences: entropy is non-increasing under operations such as conditioning, discarding a subsystem does not increase mutual information, and quantum operations (CPTP maps) do not increase mutual information, see, e.g. [20, §11.4.2] for a discussion.

2.9   Entanglement and LOCC

We have seen several ways that *entanglement* is a resource already for the space $\mathcal{H}_A \otimes \mathcal{H}_B = \mathbb{C}^2 \otimes \mathbb{C}^2$: given a shared $|epr\rangle = \frac{1}{\sqrt{2}}(|00\rangle + |11\rangle)$, one can transport two bits of classical information using only one qubit ("super dense coding") and one can also transmit one qubit of quantum information from Alice to Bob by sending two classical bits ("teleportation").

Given a quantum state $\rho \in \text{End}(\mathcal{H}_A \otimes \mathcal{H}_B)$, one would like to know how "entangled" it is, e.g., what quantum states could it be used to transport with the aid of classical communication (as in teleportation)? In this section I discuss measures of "quality of entanglement".

2.9.1 LOCC

Assume several different laboratories can communicate classically, have prepared some shared states in advance, and can perform unitary and projection operations on their parts of the states, as was the situation for quantum teleportation. More precisely, make the following assumptions:

- $\mathcal{H} = \mathcal{H}_1 \otimes \cdots \otimes \mathcal{H}_n$, and the $\mathcal{H}_j$ share an entangled state $|\psi\rangle$. Often one will just have $\mathcal{H} = \mathcal{H}_A \otimes \mathcal{H}_B$ and $|\psi\rangle = \alpha|00\rangle + \beta|11\rangle$.
- The laboratories can communicate classically.
- Each laboratory is allowed to perform unitary and measurement operations on their own spaces.

The above assumptions are called *LOCC* for "local operations and classical communication". It generalizes the set-up for teleportation Sect. 2.3.2.

Restrict to the case $\mathcal{H} = \mathcal{H}_A \otimes \mathcal{H}_B$, each of dimension two. I will use $|epr\rangle$ as a benchmark for measuring the quality of entanglement.

We will not be concerned with a single state $|\psi\rangle$, but the tensor product of many copies of it, $|\psi\rangle^{\otimes n} \in (\mathcal{H}_A \otimes \mathcal{H}_B)^{\otimes n}$. "How much" entanglement does $|\psi\rangle^{\otimes n}$ have? An answer is given in Sect. 2.9.4.

To gain insight as to which states can be produced via LOCC from a given density operator, return to the classical case. For the classical cousin of LOCC, by considering diagonal density operators, we see we should allow alteration of a probability distribution by permuting the p_j (permutation matrices are unitary), and more generally averaging our probability measure under some probability measure on elements of $\mathfrak{S}_d$ (the classical cousin of a projective measurement), i.e., we should allow

$$\overline{p} \mapsto \sum_{\sigma \in \mathfrak{S}_d} q_\sigma \mu(\sigma)\overline{p} \tag{2.18}$$

where $\mu : \mathfrak{S}_d \to GL_d$ is map sending a permutation to a $d \times d$ permutation matrix, and q is a probability distribution on $\mathfrak{S}_d$.

This is because the unitary and projection local operators allowed amount to

$$\rho \mapsto \sum_{j=1}^{k} p_j U_j \rho U_j^{-1}$$

where the U_j are unitary and p is a probability distribution on $\{1, \ldots, k\}$ for some finite k.

2.9.2 A Partial Order on Probability Distributions Compatible with Entropy

Shannon entropy is non-increasing under an action of the form (2.18). The partial order on probability distributions determined by (2.18) is the *dominance order*:

Definition 2.9.1 Let $x, y \in \mathbb{R}^d$, write $x^\downarrow$ for x re-ordered such that $x_1 \geq x_2 \geq \cdots \geq x_d$. Write $x \prec y$ if for all $k \leq d$, $\sum_{j=1}^k x_j^\downarrow \leq \sum_{j=1}^k y_j^\downarrow$.

Note that if p is a probability distribution concentrated at a point, then $\overline{q} \prec \overline{p}$ for all probability distributions q, and if p is such that $p_j = \frac{1}{d}$ for all j, then $\overline{p} \prec \overline{q}$ for all q, and more generally the dominance order is compatible with the entropy in the sense that $\overline{p} \prec \overline{q}$ implies $H(\overline{p}) \geq H(\overline{q})$.

Recall that a matrix $D \in Mat_{d \times d}$ is doubly stochastic if $D_{ij} \geq 0$ and all column and row sums equal one. Let $\mathcal{DS}_d \subset Mat_{d \times d}$ denote the set of doubly stochastic matrices. Birkhoff [5] showed $\mathcal{DS}_d$ is the convex hull of $\mu(\mathfrak{S}_d)$, and Hardy-Littlewood-Polya [13] showed $\{x \mid x \prec y\} = \mathcal{DS}_d \cdot y$.

2.9.3 A Reduction Theorem

The study of LOCC is potentially unwieldy because there can be numerous rounds of local operations and classical communication, making it hard to model. The following result eliminates this problem:

Proposition 2.9.2 *If* $|\psi\rangle \in \mathcal{H}_A \otimes \mathcal{H}_B$ *can be transformed into* $|\phi\rangle$ *by LOCC, then it can be transformed to* $|\phi\rangle$ *by the following sequence of operations:*

(1) *Alice performs a single measurement with operators* Π_{M_j}.
(2) *She sends the result of her measurement (some j) to Bob classically.*
(3) *Bob performs a unitary operation on his system.*

The key point is that for any vector spaces V, W, an element $f \in V \otimes W$, may be considered as a linear map $W^* \to V$. In our case, $\mathcal{H}_B^* \simeq \mathcal{H}_B$ so $|\psi\rangle$ induces a linear map $\mathcal{H}_B \to \mathcal{H}_A$ which gives us the mechanism to transfer Bob's measurements to Alice.

For $X \in \mathcal{H}_A \otimes \mathcal{H}_B$, let singvals($X$) denote the set of its singular values Now I can state the main theorem on LOCC:

Theorem 2.9.3 ([19]) *For states* $|\psi\rangle, |\phi\rangle \in \mathcal{H}_A \otimes \mathcal{H}_B$, $|\psi\rangle \rightsquigarrow |\phi\rangle$ *by LOCC if and only if* singvals($|\psi\rangle$) $\prec$ singvals($|\phi\rangle$).

2.9.4 Entanglement Distillation (Concentration) and Dilution

To compare the entanglement resources of two states $|\phi\rangle$ and $|\psi\rangle$, consider $|\phi\rangle^{\otimes m}$ for large m with the goal of determining the largest $n = n(m)$ such that $|\phi\rangle^{\otimes m}$ may be degenerated to $|\psi\rangle^{\otimes n}$ via LOCC. Due to the approximate and probabilistic nature of quantum computing, relax this to degenerating $|\phi\rangle^{\otimes m}$ to a state that is close to $|\psi\rangle^{\otimes n}$.

There is a subtlety for this question worth pointing out. Teleportation was defined in such a way that Alice did not need to know the state she was teleporting, but for distillation and dilution, she will need to know that its right singular vectors are standard basis vectors. More precisely, if she is in possession of $|\psi\rangle = \sqrt{p_1}|v_1\rangle \otimes |1\rangle + \sqrt{p_2}|v_2\rangle \otimes |2\rangle$, she can teleport the second half of it to Bob if they share $|epr\rangle \in \mathcal{H}_A \otimes \mathcal{H}_B$. More generally, if she is in possession of $|\psi\rangle = \sum_{j=1}^d \sqrt{p_j}|v_j\rangle \otimes |j\rangle \in \mathcal{H}_{A'} \otimes \mathcal{H}_{A''}$, she can teleport it to Bob if they share enough EPR states. In most textbooks, Alice is assumed to possess states whose singular vectors are $|jj\rangle$'s and I will follow that convention here. Similarly, if $|\psi\rangle = \sum_{j=1}^d \sqrt{p_j}|jj\rangle \in \mathcal{H}_A \otimes \mathcal{H}_B$, I discuss how many shared EPR states they can construct from a shared $|\psi\rangle^{\otimes m}$.

Define the *entanglement cost* $E_C(\psi)$ to be $\inf_m \frac{n(m)}{m}$ where $n(m)$ copies of ψ can be constructed from $|epr\rangle^{\otimes m}$ by LOCC with error going to zero as $m \to \infty$. Similarly, define the *entanglement value*, or *distillable entanglement* $E_V(\psi)$ to be $\sup_m \frac{n(m)}{m}$ where $n(m)$ copies of $|epr\rangle$ can be constructed with diminishing error from $|\psi\rangle^{\otimes m}$ by LOCC. One has $E_V(\psi) = E_C(\psi) = H(|\psi\rangle\langle\psi|)$.

Remark 2.9.4 In classical computation one can reproduce information, but this cannot be done with quantum information in general. This is because the map $|\psi\rangle \mapsto |\psi\rangle \otimes |\psi\rangle$, called the *Veronese map* in algebraic geometry, is not a linear map. This observation is called the *no cloning theorem* in the quantum literature. However, one can define a linear map, e.g., $\mathbb{C}^2 \to \mathbb{C}^2 \otimes \mathbb{C}^2$ that duplicates basis vectors, i.e., $|0\rangle \mapsto |0\rangle \otimes |0\rangle$ and $|1\rangle \mapsto |1\rangle \otimes |1\rangle$. But then of course $\alpha|0\rangle + \beta|1\rangle \mapsto \alpha|0\rangle \otimes |0\rangle + \beta|1\rangle \otimes |1\rangle \neq (a|0\rangle + \beta|1\rangle)^{\otimes 2}$.

For mixed states ρ on $\mathcal{H}_A \otimes \mathcal{H}_B$, one can still define $E_C(\rho)$ and $E_V(\rho)$, but there exist examples where they differ, so there is not a canonical measure of entanglement. A wish list of what one might want from an entanglement measure E:

- Non-increasing under LOCC.
- If ρ is a product state, i.e., $\rho = |\phi_A\rangle\langle\phi_A| \otimes |\psi_B\rangle\langle\psi_B|$, then $E(\rho) = 0$.

The two conditions together imply any state constructible from a product state by LOCC should also have zero entanglement. Hence the following definition:

Definition 2.9.5 A density operator $\rho \in \text{End}(\mathcal{H}_1 \otimes \cdots \otimes \mathcal{H}_n)$ is *separable* if $\rho = \sum_i p_i \rho_{i,1} \otimes \cdots \otimes \rho_{i,n}$, where $\rho_{i,\alpha} \in \text{End}(\mathcal{H}_\alpha)$ are density operators, $p_i \geq 0$, and $\sum_i p_i = n$. If ρ is not separable, ρ is *entangled*.

Definition 2.9.6 An *entanglement monotone* E is a function on density operators on $\mathcal{H}_A \otimes \mathcal{H}_B$ that is non-increasing under LOCC.

An example of an entanglement monotone different from E_V, E_C useful for general density operators is the *squashed entanglement* [9]

$$E_{sq}(\rho_{AB}) := \inf_C \{\frac{1}{2}[H(\rho_A \mid \rho_C) + H(\rho_B \mid \rho_C) - H(\rho_{AB} \mid \rho_C)] \mid \rho_{AB} = \text{trace}_{\mathcal{H}_C}(\rho_{ABC})\}.$$

For bipartite states, all entanglement measures are compatible with the order of states from most to least entangled. This breaks down already for tripartite states.

Remark 2.9.7 An entanglement measure appealing to geometers is SLOCC (stochastic local operations and classical communication) defined originally in [4], which asks if $|\psi\rangle \in \mathcal{H}_1 \otimes \cdots \otimes \mathcal{H}_d$ is in the same $SL(\mathcal{H}_1) \times \cdots \times SL(\mathcal{H}_d)$ orbit as $|\phi\rangle \in \mathcal{H}_1 \otimes \cdots \otimes \mathcal{H}_d$. If one relaxes this to orbit closure, then it amounts to being able to convert $|\psi\rangle$ to $|\phi\rangle$ with positive probability. While appealing, and while there is literature on SLOCC, given the probabilistic nature of quantum computing, its use appears to be limited to very special cases, where the orbit structure is understood (e.g., $d \leq 4$, $\dim \mathcal{H}_j = 2$).

2.10 Tensor Network States

Assuming interactions between particles should be short-ranged enough (which is satisfied in most physically relevant set-ups), if we have an arrangement of electrons, say on a circle, as in Fig. 2.3.

It is highly improbable that the electrons will share entanglement with any but their nearest neighbors. This is fortuitous, because if one is dealing with thousands of electrons and would like to describe their joint state, a priori one would have to work with a vector space of dimension 2^n, with n in the thousands, which is not

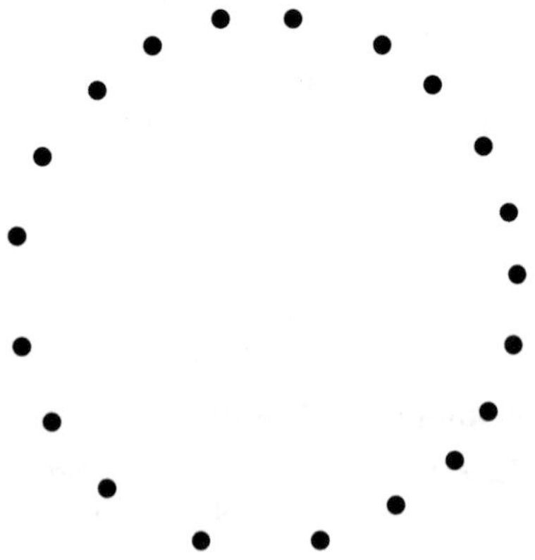

Fig. 2.3 Electrons arranged on a circle

feasible. The practical solution to this problem is to define a subset of $(\mathbb{C}^2)^{\otimes n}$ of reasonable dimension (e.g. $O(n)$) consisting of the probable states.

For another example, say the isolated system consists of electrons arranged along a line as below.

and we only want to allow electrons to be entangled with their nearest neighbors. This leads to the notion of *Matrix Product States (MPS)*: draw a graph reflecting this geometry, with a vertex for each electron. To each vertex, attach edges going from the electron's vertex to those of its nearest neighbors, and add an additional edge not attached to anything else (these will be called physical edges). If our space is $\mathcal{H}_1 \otimes \cdots \otimes \mathcal{H}_n$, then, assuming vertex j has two neighbors, attach two auxiliary vector spaces, E_{j-1}, E_j^*, and a tensor $T_j \in \mathcal{H}_j \otimes E_{j-1} \otimes E_j^*$. If we are on a line, to vertex one, we just attach $T_1 \in \mathcal{H}_1 \otimes E_1^*$, and similarly, to vertex n we attach $T_n \in \mathcal{H}_n \otimes E_{n-1}$. Now consider the tensor product of all the tensors

$$T_1 \otimes \cdots \otimes T_n \in (\mathcal{H}_1 \otimes E_1^*) \otimes (\mathcal{H}_2 \otimes E_1 \otimes E_2^*) \otimes \cdots \otimes (\mathcal{H}_{n-1} \otimes E_{n-2} \otimes E_{n-1}^*) \otimes (\mathcal{H}_n \otimes E_{n-1})$$

Assume each E_j has dimension k. We can contract these to obtain a tensor $T \in \mathcal{H}_1 \otimes \cdots \otimes \mathcal{H}_n$. If $k = 1$, we just obtain the product states. As we increase k, we obtain a steadily larger subset of $\mathcal{H}_1 \otimes \cdots \otimes \mathcal{H}_n$, that fills the entire space for sufficiently large (exponential size) k. The claim is that the tensors obtainable in this fashion (for some k determined by the physical setup) are exactly those locally entangled states that we seek. (The first and last tensors in this set-up may be interpreted as boundary values, related to interaction with the outside world.)

For the circle, the only difference in the construction is to make the construction periodic, so $T_1 \in \mathcal{H}_1 \otimes E_n \otimes E_1^*$ and $T_n \in \mathcal{H}_n \otimes E_{n-1} \otimes E_n^*$. Such states are called *Matrix product states* or *MPS* in the physics literature.

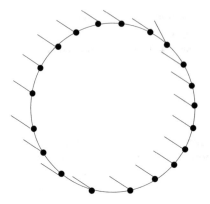

Sometimes for applications (e.g. translation invariant systems on the circle) one requires the same tensor be placed at each vertex. If the tensor is $\sum_{i,j,\alpha} T_{i,j,\alpha} \langle i| \otimes |j\rangle \otimes v_\alpha$, the resulting tensor is $\sum T_{i_1,i_2,\alpha_1} T_{i_2,i_3,\alpha_2} \cdots T_{i_n,i_1,\alpha_n} v_{\alpha_1} \otimes \cdots \otimes v_{\alpha_n}$.

For a second example, consider electrons arranged in a rectangular array (or on a grid on a torus), where each vertex is allowed to interact with its four nearest neighbors. Such states are called *projected entangled pair states* or *PEPS* in the physics literature.

Assume we place the same tensor at each vertex. If our grid is $n \times n$ and periodic, we obtain a map $(\mathbb{C}^k)^{\otimes 4} \otimes \mathbb{C}^d \to (\mathbb{C}^d)^{\otimes n^2}$.

Definition 2.10.1 Let Γ be a directed graph with vertices v_α and two kinds of edges: "physical" edges e_i, that are attached to a single vertex, and "auxiliary" (or *entanglement*) edges e_s between two vertices. Associate to each physical edge a vector space V_i (in the quantum case, $V_i = \mathcal{H}_i$ is a Hilbert space), and to each auxiliary edge a vector space E_s, of dimension $\mathbf{e}_s$. Let $\bar{\mathbf{e}} = (\mathbf{e}_1, \ldots, \mathbf{e}_f)$ denote the vector of these dimensions. A *tensor network state* associated to $(\Gamma, \{V_i\}, \bar{\mathbf{e}})$ is a tensor $T \in V_1 \otimes \cdots \otimes V_n$ obtained as follows: To each vertex v_α, associate a tensor

$$T_\alpha \in \otimes_{i \in \alpha} V_i \otimes_{s \in in(\alpha)} E_s^* \otimes_{t \in out(\alpha)} E_t.$$

Here $in(\alpha)$ are the edges going into vertex α and $out(\alpha)$ are the edges going out of the vertex. The *tensor network state* associated to this configuration is $T := contr(T_1 \otimes \cdots \otimes T_g) \in V_1 \otimes \cdots \otimes V_n$. Let $TNS(\Gamma, V_1 \otimes \cdots \otimes V_n, \mathbf{e}) \subset V_1 \otimes \cdots \otimes V_n$ denote the set of tensor network states.

Other graphs that occur are trees, which also appear in the numerical analysis literature, see [12].

Example 2.10.2 Let Γ be:

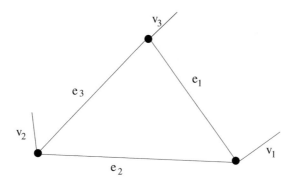

Then

$$TNS(\Gamma, V_1 \otimes V_2 \otimes V_3, \overline{e}) = TNS(\Gamma, (E_1^* \otimes E_2) \otimes (E_2^* \otimes E_3) \otimes (E_3^* \otimes E_1), \overline{e})$$
$$= \text{End}(V_1) \times \text{End}(V_2) \times \text{End}(V_3) \cdot M_{\langle e_1, e_2, e_3 \rangle}.$$

Here $M_{\langle e_1, e_2, e_3 \rangle}$ is the *matrix multiplication tensor*, for $A \in Mat_{e_1 \times e_2}$, $B \in Mat_{e_2 \times e_3}$, $C \in Mat_{e_3 \times e_1}$, $(A, B, C) \mapsto \text{trace}(ABC)$. Let $e_1, \ldots, e_{e_1}$ be a basis of E_1, $f_1, \ldots, f_{e_2}$ be a basis of E_2, and $g_1, \ldots, g_{e_3}$ be a basis of E_3.

There are many open questions about tensor network states: only in very few cases is there a satisfactory description of the states producible from a given graph and parameters. Regarding algebraic geometry, one can ask for a description of the ideal of the Zariski closure of the set of states producible from a given graph and parameters. Such information would be extremely useful for applications.

2.11   Representation Theory in Quantum Information Theory

Say we have a state $\rho \in \mathcal{H}_A \otimes \mathcal{H}_B$ or in $\mathcal{H}_{A_1} \otimes \cdots \otimes \mathcal{H}_{A_d}$ create-able by a device or experiment and we perform the experiment numerous times to get a state $\rho^{\otimes n} \in \mathcal{H}_A^{\otimes n} \otimes \mathcal{H}_B^{\otimes n}$. What is the "value" of such states for information theory? What are measurements of these states likely to produce? It turns out (partial) answers to these questions can be gained by exploiting representation theory, see Theorem 2.11.5. I review the relevant representation theory and then apply it to describe the solution to the quantum marginal problem, which discusses which pairs of states on $\mathcal{H}_A$, $\mathcal{H}_B$ may arise as partial traces of some $\rho_{AB} \in \text{End}(\mathcal{H}_A \otimes \mathcal{H}_B)$.

2.11.1   Review of Relevant Representation Theory

(Isomorphism classes of) irreducible representations of the permutation group $\mathfrak{S}_d$ are indexed by partitions of d, write $[\pi]$ for the $\mathfrak{S}_d$-module corresponding to the partition π. The irreducible polynomial representations of $GL(V)$ are indexed by partitions $\pi = (p_1, \ldots, p_{\ell(\pi)})$ with $\ell(\pi) \leq \dim V$. Write $S_\pi V$ for the corresponding $GL(V)$-module.

Theorem 2.11.1 (Schur-Weyl Duality) *As a $GL(V) \times \mathfrak{S}_d$-module,*

$$V^{\otimes d} = \bigoplus_{|\pi|=d} S_\pi V \otimes [\pi].$$

Let $P_\pi : V^{\otimes d} \to S_\pi V \otimes [\pi]$ denote the $GL(V) \times \mathfrak{S}_d$-module projection operator.

One is often interested in decompositions of a module under the action of a subgroup. For example $S^d(V \otimes W)$ is an irreducible $GL(V \otimes W)$-module, but as a $GL(V) \times GL(W)$-module it has the decomposition, called the *Cauchy formula*,

$$S^d(V \otimes W) = \oplus_{|\pi|=d} S_\pi V \otimes S_\pi W. \tag{2.19}$$

We will be particularly interested in the decomposition of $S^d(U \otimes V \otimes W)$ as a $GL(U) \times GL(V) \times GL(W)$-module. An explicit formula for this decomposition is *not known*. Write

$$S^d(U \otimes V \otimes W) = \bigoplus_{|\pi|,|\mu|,|\nu|=d} (S_\pi U \otimes S_\mu V \otimes S_\nu W)^{\oplus k_{\pi,\mu,\nu}}.$$

The numbers $k_{\pi,\nu,\mu}$ that record the multiplicities are called *Kronecker coefficients*. They have several additional descriptions. For example, $S_\pi(V \otimes W) = \bigoplus_{|\mu|,|\nu|=d}(S_\mu V \otimes S_\nu W)^{\oplus k_{\pi,\mu,\nu}}$, and $k_{\pi,\mu,\nu} = \dim([\pi] \otimes [\mu] \otimes [\mu])^{\mathfrak{S}_d} = \mathrm{mult}([d], [\pi] \otimes [\mu] \otimes [\nu]) = \mathrm{mult}([\pi], [\mu] \otimes [\nu])$.

2.11.2   Quantum Marginals and Projections onto Isotypic Subspaces of $\mathcal{H}^{\otimes d}$

In this section I address the question: what are compatibility conditions on density operators ρ on $\mathcal{H}_A \otimes \mathcal{H}_B$, ρ' on $\mathcal{H}_A$ and ρ'' on $\mathcal{H}_B$ such that $\rho' = \mathrm{trace}_{\mathcal{H}_B}(\rho)$, $\rho'' = \mathrm{trace}_{\mathcal{H}_A}(\rho)$? As you might expect by now, compatibility will depend only on the spectra of the operators.

Above I discussed representations of the general linear group $GL(V)$ where V is a complex vector space. In quantum theory, one is interested in representations on the unitary group $\mathbf{U}(\mathcal{H})$ on a Hilbert space $\mathcal{H}$. The unitary group is a real Lie group, not a complex Lie group, because complex conjugation is not a complex linear map. It is a special case of a general fact about representations of a maximal compact subgroups of complex Lie groups have the same representation theory as the original group, so in particular the decomposition of $\mathcal{H}^{\otimes d}$ as a $\mathbf{U}(\mathcal{H})$-module coincides with its decomposition as a $GL(\mathcal{H})$-module.

For a partition $\pi = (p_1, \ldots, p_d)$ of d, introduce the notation $\overline{\pi} = (\frac{p_1}{d}, \ldots, \frac{p_d}{d})$ which is a probability distribution on $\{1, \ldots, d\}$.

Theorem 2.11.2 ([8])   *Let ρ_{AB} be a density operator on $\mathcal{H}_A \otimes \mathcal{H}_B$. Then there exists a sequence (π_j, μ_j, ν_j) of triples of partitions such that $k_{\pi_j, \mu_j, \nu_j} \neq 0$ for all j and*

$$\lim_{j \to \infty} \overline{\pi}_j = \mathrm{spec}(\rho_{AB})$$

$$\lim_{j \to \infty} \overline{\mu}_j = \mathrm{spec}(\rho_A)$$

$$\lim_{j \to \infty} \overline{\nu}_j = \mathrm{spec}(\rho_B).$$

Theorem 2.11.3 ([17])   *Let ρ_{AB} be a density operator on $\mathcal{H}_A \otimes \mathcal{H}_B$ such that $\mathrm{spec}(\rho_{AB})$, $\mathrm{spec}(\rho_A)$ and $\mathrm{spec}(\rho_B)$ are all rational vectors. Then there exists an integer $M > 0$ such that*

$$k_{M \, \mathrm{spec}(\rho_A), M \, \mathrm{spec}(\rho_B), M \, \mathrm{spec}(\rho_C)} \neq 0.$$

Theorem 2.11.4 ([17])   *Let π, μ, ν be partitions of d with $k_{\pi, \mu, \nu} \neq 0$ and satisfying $\ell(\pi) \leq mn$, $\ell(\mu) \leq m$, and $\ell(\nu) \leq n$. Then there exists a density operator ρ_{AB} on $\mathbb{C}^n \otimes \mathbb{C}^m = \mathcal{H}_A \otimes \mathcal{H}_B$ with $\mathrm{spec}(\rho_{AB}) = \overline{\pi}$, $\mathrm{spec}(\rho_A) = \overline{\mu}$, and $\mathrm{spec}(\rho_B) = \overline{\nu}$.*

Klyatchko's proofs are via co-adjoint orbits and vector bundles on flag varieties, while the proof of Christandl-Mitchison is information-theoretic in flavor.

Recall the relative entropy $H(\overline{p}||\overline{q}) = -\sum_i p_i \log \frac{q_i}{p_i}$, which may be thought of as measuring how close p, q are because it is non-negative, and zero if and only if $p = q$. A key step in the Christandl-Mitchison proof is the following theorem:

Theorem 2.11.5 ([14])   *Let $\rho \in \mathrm{End}(\mathcal{H})$ be a density operator, where $\dim \mathcal{H} = n$. Let $|\pi| = d$ and let $P_\pi : \mathcal{H}^{\otimes d} \to S_\pi \mathcal{H}[\pi]$ be the projection operator. Then*

$$\mathrm{trace}(P_\pi \rho^{\otimes d}) \leq (d+1)^{\binom{n}{2}} e^{-d H(\overline{\pi} || \, \mathrm{spec}(\rho))}.$$

A key step of the proof is that the projection of e_I to $S_\pi V \otimes [\pi]$ is nonzero if and only if $wt(e_I) \prec \pi$.

Let $Spec_{m,n,mn}$ denote the set of admissible triples $(\mathrm{spec}(\rho_A), \mathrm{spec}(\rho_B), \mathrm{spec}(\rho_{AB}))$ and $KRON_{m,n,mn}$ the triples $(\overline{\mu}, \overline{\nu}, \overline{\pi})$ of normalized partitions (μ, ν, π) with $\ell(\mu) \leq m$, $\ell(\nu) \leq n$, $\ell(\pi) \leq mn$ and $k_{\pi, \mu, \nu} \neq 0$.

The theorems above imply:

$$Spec_{m,n,mn} = \overline{KRON_{m,n,mn}}.$$

In particular, $Spec_{m,n,mn}$ is a convex polytope.

Acknowledgements I thank the organizers of the International workshop on Quantum Physics and Geometry, especially Alessandra Bernardi, who also co-organized an intensive summer class on Quantum computing and quantum information theory that I gave June–July 2017. I also thank L. Chiantini, F. Gesmundo, F. Holweck, and G. Ottaviani for useful comments on a draft of this article. I am especially grateful to the anonymous referee for a very careful reading of the draft and numerous useful suggestions.

The author Landsberg supported by NSF grant DMS-1405348.

References

1. S. Aaronson, *Quantum Computing Since Democritus* (Cambridge University Press, Cambridge, 2013). MR 3058839
2. S. Arora, B. Barak, *Computational Complexity: A Modern Approach* (Cambridge University Press, Cambridge, 2009). MR 2500087 (2010i:68001)
3. J.S. Bell, On the Einstein-Podolsky-Rosen paradox. Physics **1**, 195–200 (1964)
4. C.H. Bennett, S. Popescu, D. Rohrlich, J.A. Smolin, A.V. Thapliyal, Exact and asymptotic measures of multipartite pure-state entanglement. Phys. Rev. A **63**, 012307 (2000)
5. G. Birkhoff, Three observations on linear algebra. Univ. Nac. Tucumán Rev. A. **5**, 147–151 (1946). MR 0020547
6. F.G.S.L. Brandao, M. Christandl, A.W. Harrow, M. Walter, The Mathematics of Entanglement, ArXiv e-prints (2016)
7. J.F. Clauser, M.A. Horne, A. Shimony, R.A. Holt, Proposed experiment to test local hidden-variable theories. Phys. Rev. Lett. **23**, 880–884 (1969)
8. M. Christandl, G. Mitchison, The spectra of quantum states and the Kronecker coefficients of the symmetric group. Commun. Math. Phys. **261**(3), 789–797 (2006). MR 2197548
9. M. Christandl, A. Winter, "Squashed entanglement": an additive entanglement measure. J. Math. Phys. **45**(3), 829–840 (2004). MR 2036165
10. A. Einstein, B. Podolsky, N. Rosen, Can quantum-mechanical description of physical reality be considered complete? Phys. Rev. **47**, 777–780 (1935)
11. P. Erdös, Some remarks on the theory of graphs. Bull. Am. Math. Soc. **53**, 292–294 (1947). MR 0019911
12. W. Hackbusch, *Tensor Spaces and Numerical Tensor Calculus*. Springer Series in Computational Mathematics, vol. 42 (Springer, Heidelberg, 2012). MR 3236394
13. G.H. Hardy, J.E. Littlewood, G. Pólya, *Inequalities*, 2nd edn. (Cambridge University Press, Cambridge, 1952). MR 0046395
14. M. Keyl, R.F. Werner, Estimating the spectrum of a density operator. Phys. Rev. A (3) **64**(5), 052311 (2001). MR 1878924
15. A.Yu. Kitaev, A.H. Shen, M.N. Vyalyi, *Classical and Quantum Computation*. Graduate Studies in Mathematics, vol. 47 (American Mathematical Society, Providence, 2002). Translated from the 1999 Russian original by Lester J. Senechal. MR 1907291
16. O. Klein, Quantum coding. Z. Phys. **72**, 767–775 (1931)
17. A. Klyachko, Quantum marginal problem and representations of the symmetric group (2004). Preprint, arXiv:quant-ph/0409113v1

18. J.M. Landsberg, Quantum computation and information: Notes for fall 2017 TAMU class (2017). Available at http://www.math.tamu.edu/~jml/quantumnotes.pdf
19. M.A. Nielsen, Conditions for a class of entanglement transformations. Phys. Rev. Lett. **83**, 436–439 (1999)
20. M.A. Nielsen, I.L. Chuang, *Quantum Computation and Quantum Information* (Cambridge University Press, Cambridge, 2000). MR MR1796805 (2003j:81038)
21. B. Schumacher, Quantum coding. Phys. Rev. A (3) **51**(4), 2738–2747 (1995). MR 1328824
22. C.E. Shannon, A mathematical theory of communication. Bell Syst. Tech. J. **27**, 379–423, 623–656 (1948). MR 0026286

Chapter 3
Entanglement, CP-Maps and Quantum Communications

Davide Pastorello

Abstract In this chapter we review the employment of quantum entanglement as a resource for information processing and transmission. In particular we introduce and discuss the notion of *completely positive maps* operating on observable algebras of physical systems in order to have a model to construct communication channels based on quantum processes. Then we discuss advantages and limitations of entanglement-assisted quantum communication schemes like quantum teleportation and dense coding.

3.1 Introduction

In this introductory section we have a look on fundamentals of quantum mechanics in order to discuss the notion of quantum entanglement and its properties in the second section. The third section is devoted to introducing the formalism of quantum channels as completely-positive maps (CP-maps) that is the framework where we describe the *impossible machines* (i.e. a collection of no-go theorems) and the quantum communication schemes in fourth section. In the last section there is a final comment on future perspectives about some topics of the work.

The standard formulation of quantum mechanics (QM) prescribes that a complex Hilbert space H (finite-dimensional or infinite-dimensional and separable) is associated to any quantum system, let us denote the inner product of H by $\langle \ | \ \rangle$. Let $\mathfrak{B}(H)$ and $\mathfrak{B}_1(H) \subset \mathfrak{B}(H)$ be the spaces of bounded operators and trace class operators[1]

---

[1] Let us recall that $T \in \mathfrak{B}(H)$ is called *trace class operator* if $\sum_{\psi \in N} \langle \psi | \ |T| \ \psi \rangle < \infty$ for some complete orthonormal system $N \subset H$, where $|T| := \sqrt{T^\dagger T}$ and $T^\dagger$ denotes the adjoint of T.

D. Pastorello (✉)
Department of Mathematics, University of Trento, Trento Institute for Fundamental Physics and Applications (TIFPA), Povo, Trento, Italy
e-mail: d.pastorello@unitn.it

© Springer Nature Switzerland AG 2019
E. Ballico et al. (eds.), *Quantum Physics and Geometry*,
Lecture Notes of the Unione Matematica Italiana 25,
https://doi.org/10.1007/978-3-030-06122-7_3

on H respectively. The set of *quantum states* is given by the positive trace class operators on H with unit trace:

$$\mathfrak{S}(H) = \{\rho \in \mathfrak{B}_1(H) : \rho \geq 0, \mathrm{tr}(\rho) = 1\}. \tag{3.1}$$

This set is convex in $\mathfrak{B}_1(H)$ and its extreme elements are called *pure states*, they correspond to the orthogonal projectors of rank 1 in H, so a pure state can be written[2] as $\rho = |\psi\rangle\langle\psi|$ for some $\psi \in H$ with $\|\psi\| = 1$. According to Krein-Millman theorem any quantum state can be decomposed as a convex combination of pure states with a natural statistical interpretation; The non-pure states are called *mixed states*.

Time evolution of an isolated quantum system is described by a continuous one-parameter group of unitary operators $\{U(t)\}_{t \in \mathbb{R}^+}$ on the Hilbert space H of the theory. If $\rho_1 \in \mathfrak{S}(H)$ is the state of the system at time t_1 and $\rho_2 \in \mathfrak{S}(H)$ is the state of the system at time $t_2 > t_1$ then:

$$\rho_2 = U(t_2 - t_1)\rho_1 U^\dagger(t_2 - t_1). \tag{3.2}$$

The *measurement processes* on quantum systems are represented by *positive operators valued measures*[3] (POVMs). In particular if a measurement process admits a finite set $X \subset \mathbb{R}$ of possible experimental outcomes then it is represented by a POVM that is nothing but a finite collection of positive operators $M = \{E_x\}_{x \in X} \subset \mathfrak{B}(H)$ such that $\sum_{x \in X} E_x = \mathbb{I}$. The physical interpretation of the POVM M is the following: The probability to obtain the outcome $x \in X$ by the considered measurement process performed on the system in the state $\rho \in \mathfrak{S}(H)$ is:

$$\mathfrak{P}_\rho^M(x) = \mathrm{tr}(E_x \rho). \tag{3.3}$$

The building blocks of measurement processes, i.e. the operators $E \in \mathfrak{B}(H)$ such that $0 \leq E \leq \mathbb{I}$, are called *effects*. When the effects of a POVM are orthogonal projectors $\{P_k\}_k$ then the notion of *post-measurement state* can be defined as following: If the measurement process $\{P_k\}_k$ is performed on a quantum system in the state ρ producing the outcome k then the state of the system is mapped in:

$$\rho_k = \frac{P_k \rho P_k}{\mathrm{tr}(P_k \rho)}. \tag{3.4}$$

When the measurement is described by a general POVM $\{E_k\}_k$ a notion of post-measurement state can be defined as well but it requires further information. In fact

---

However we will focus on finite-dimensional Hilbert spaces where every linear operator is bounded and trace class.

[2] The symbol $|\psi\rangle\langle\varphi|$ denotes the outer product of vectors ψ and φ.

[3] See [16] for a complete description of POVMs as a generalization of spectral measures.

we need a decomposition of any effect into *measuring operators* $E_k = M_k^\dagger M_k$ so that the post-measurement state is:

$$\rho_k = \frac{M_k \rho M_k^\dagger}{\mathrm{tr}(M_k \rho)}. \tag{3.5}$$

Since M_k are not required to be positive, we have an infinite number of operators satisfying $E_k = M_k^\dagger M_k$ for a given E_k. From the physical viewpoint there are infinite experimental apparatuses giving the same outcome statistic.

The mapping $\rho \mapsto \rho_k$ is historically called *wave function collapse* and represents a slightly naive[4] way to describe the aftermath of a measurement process on the measured system in QM.

From a slightly different viewpoint, the main mathematical object to describe a quantum system is the C^*-algebra[5] $\mathfrak{B}(\mathsf{H})$ of bounded operators on a separable (or finite-dimensional) Hilbert space H, where the involution is given by the adjoint operation and the norm is:

$$\| B \| := \sup_{\|\psi\|=1} \| B\psi \|. \tag{3.6}$$

For the finite-dimensional case, $\mathfrak{B}(\mathsf{H})$ is the algebra of linear operators on the Hilbert space $\mathsf{H} \simeq \mathbb{C}^n$.

Even in classical mechanics we find a C^*-algebra which plays a central role, in fact consider a classical system described in the phase space $\mathcal{M}$ (a real symplectic manifold), classical states are given by probability densities on $\mathcal{M}$ and the physical quantities are represented by real elements of the abelian C^*-algebra $\mathscr{C}^0(\mathcal{M})$ of continuous functions vanishing at infinity on $\mathcal{M}$ equipped with the pointwise product, the uniform norm and the involution given by complex conjugation.

---

[4]Within the standard formulation, QM does not establish what a measuring device is (it only assumes its existence) and it is not capable of describing the interaction between an instrument and a quantum system beyond the notion of post-selection mapping.

[5]A linear associative algebra $\mathfrak{A}$ on $\mathbb{C}$ is called C^*-*algebra* if it satisfies the following requirements:

1. $\mathfrak{A}$ is a Banach algebra, i.e. it is a normed space such that its norm $\| \ \|$ satisfies:
   $\| AB \| \leq \| A \| \| B \|$ for all $A, B \in \mathfrak{A}$ and $\mathfrak{A}$ is complete w.r.t. the topology induced by $\| \ \|$;
2. $\mathfrak{A}$ is equipped with an involution $\dagger : \mathfrak{A} \to \mathfrak{A}$ such that:

$$(A + B)^\dagger = A^\dagger + B^\dagger \quad (\lambda A)^\dagger = \bar{\lambda} A^\dagger \quad (AB)^\dagger = B^\dagger A^\dagger \quad (A^\dagger)^\dagger = A;$$

   for all $A, B \in \mathfrak{A}$ and $\lambda \in \mathbb{C}$;
3. $\| A^\dagger A \| = \| A \|^2$ for all $A \in \mathfrak{A}$.

Thus we assume[6] that any physical system is described in a C^*-algebra $\mathfrak{A}$ with unit $\mathbb{I}$, called *observable algebra*, where the set of states $\mathfrak{S}(\mathfrak{A})$ and the set of effects $\mathfrak{E}(\mathfrak{A})$ can be defined:

$$\mathfrak{S}(\mathfrak{A}) := \{\rho \in \mathfrak{A}^* : \rho \geq 0, \ \rho(\mathbb{I}) = 1\} \ , \quad \mathfrak{E}(\mathfrak{A}) := \{E \in \mathfrak{A} : \mathbb{I} \geq E \geq 0\}. \quad (3.7)$$

where $\mathfrak{A}^*$ denotes the dual of $\mathfrak{A}$ and the partial order relation $\geq$ in $\mathfrak{A}$ is defined as follows: $A \in \mathfrak{A}$ is said to be *positive* $(A \geq 0)$ if and only if there exists $B \in \mathfrak{A}$ such that $A = B^\dagger B$, and $A \geq B$ if and only if $A - B$ is positive.

The states are defined as linear functionals on $\mathfrak{A}$ that are positive[7] and normalized to 1, the number $\rho(E) \in [0, 1]$, for $E \in \mathfrak{E}(\mathfrak{A})$, is interpreted as the probability to obtain a fixed outcome (associated to the effect E) of a certain measurement process. The sets $\mathfrak{S}(\mathfrak{A})$ and $\mathfrak{E}(\mathfrak{A})$ are convex in $\mathfrak{A}^*$ and $\mathfrak{A}$ respectively, so the notion of pure states is natural also in this picture, on the other hand the extreme elements of $\mathfrak{E}(\mathfrak{A})$ are called *elementary propositions*.

In this work we are not interested in the abstract algebraic approach to QM, so let us consider a quantum system described in the concrete algebra $\mathfrak{B}(H)$, where $\dim H < +\infty$. Since $\mathfrak{B}(H)$ with Hilbert-Schmidt product $(A|B)_2 := \mathrm{tr}(A^\dagger B)$ is a Hilbert space, if we consider a functional $\rho \in \mathfrak{B}^*(H)$ then there exists a unique operator $\hat{\rho} \in \mathfrak{B}(H)$ such that $\rho(A) = (\hat{\rho}|A)_2 = \mathrm{tr}(\hat{\rho}A)$ for any $A \in \mathfrak{B}(H)$ with $\hat{\rho} \geq 0$ and $\mathrm{tr}(\hat{\rho}) = 1$, by Riesz representation theorem. Then we recover the identification of quantum states as operators in $\mathfrak{S}(H)$ (so-called *density matrices*) from the algebraic definition. The dual $\mathfrak{B}^*(H)$ can be identified with $\mathfrak{B}(H)$ itself of course, however we often use different notations to distinguish Schrödinger and Heisenberg pictures within the *quantum channel formalism* introduced in Sect. 3.3.

Definition 3.1 When a physical system is described in the C^*-algebra $\mathfrak{B}(H)$ with $\dim H = n$, we call it **n-level quantum system**, in particular when $n = 2$ we call it **qubit**.

An important feature in quantum information theory is the interaction between quantum and classical physical systems, now let us consider the description of classical systems. In particular we restrict to classical systems that can be described in a finite *sample space*, like a *dice* with sample space given by $X = \{1, 2, 3, 4, 5, 6\}$ or a *classical bit* with $X = \{0, 1\}$. The observable algebra of a classical system characterized by a finite sample space X is given by:

$$\mathfrak{A} = \mathscr{C}(X) = \{f : X \to \mathbb{C}\} \quad (3.8)$$

---

[6]The assumption is motivated by a list of general operational motivations to describe physical systems (classical and quantum) in abstract C^*-algebras that can be found in [23] for instance.

[7]A linear functional on $\mathfrak{A}$ is positive if $\rho(A) \geq 0 \ \forall A \geq 0$.

where the product, the involution and the C^*-norm are respectively given by:

$$(f \cdot g)(x) := f(x)g(x) \quad , \quad f^\dagger := \overline{f} \quad , \quad \| f \| := \max_X |f| \quad f, g \in \mathscr{C}(X).$$

By the *Gelfand-Naimark theorem* [23] we know that any C^*-algebra with unit is isomorphic to an algebra of bounded operators in a Hilbert space. In our finite-dimensional framework the application of such a general result is rather simple. So it may be useful representing the algebra of functions $\mathscr{C}(X)$ as an operator algebra following the approach adopted in [9] and [14]. Let us consider a Hilbert space H with dim H $= |X|$ and fix an orthonormal basis $\{e_x\}_{x \in X}$, we can define a map $\mathscr{C}(X) \ni f \mapsto \hat{f} \in \mathfrak{B}(H)$ to associate an operator to every function:

$$f \mapsto \hat{f} := \sum_{x \in X} f(x) P_x \qquad P_x := |e_x\rangle\langle e_x|. \tag{3.9}$$

Then $\mathscr{C}(X)$ can be identified with an operator algebra denoted by the same symbol. Let $\rho \in \mathscr{C}^*(X)$ be a state, let us define a map $\check{\rho}$ on X as $\check{\rho}(x) := \rho(P_x)$, so the action of ρ on the effect $\hat{f} \in \mathfrak{E}(\mathscr{C}(X))$, giving the probability to obtain a fixed outcome by a classical measurement, is:

$$\rho(\hat{f}) = \rho\left(\sum_{x \in X} f(x) P_x\right) = \sum_{x \in X} f(x)\rho(P_x) = \sum_{x \in X} f(x)\check{\rho}(x), \tag{3.10}$$

hence $\check{\rho} : X \to [0, 1]$ is a probability distribution on X, i.e. the classical notion of state indeed. Summarizing: Effects in $\mathfrak{E}(\mathscr{C}(X))$ are random variables in X and states in $\mathfrak{S}(\mathscr{C}(X))$ are probability distributions on X, then we recover classical probability theory in terms of an operator algebra.

Let $\mathfrak{A}_1 \subseteq \mathfrak{B}(H_1)$ and $\mathfrak{A}_2 \subseteq \mathfrak{B}(H_2)$ be the algebras of two physical systems S_1 and S_2 (classical or quantum). The algebra of the composite system obtained putting together S_1 and S_2 is given by the tensor product $\mathfrak{A} = \mathfrak{A}_1 \otimes \mathfrak{A}_2$. If both systems are quantum, i.e. $\mathfrak{A}_1 = \mathfrak{B}(H_1)$ and $\mathfrak{A}_2 = \mathfrak{B}(H_2)$, then we have

$$\mathfrak{A} = \mathfrak{B}(H_1) \otimes \mathfrak{B}(H_2) = \mathfrak{B}(H_1 \otimes H_2). \tag{3.11}$$

Let us assume S_1 and S_2 are both classical then $\mathfrak{A}_1 = \mathscr{C}(X_1) \subset \mathfrak{B}(H_1)$ and $\mathfrak{A}_1 = \mathscr{C}(X_2) \subset \mathfrak{B}(H_2)$ so the observable algebra of the composite system is:

$$\mathfrak{A} = \mathscr{C}(X_1 \times X_2) \subset \mathfrak{B}(H_1 \otimes H_2), \tag{3.12}$$

thus states and effects are respectively probability distributions and random variables on the Cartesian product $X_1 \times X_2$ of the sample spaces as provided by the classical probability theory.

If $\mathfrak{A}_1 = \mathscr{C}(X)$ and $\mathfrak{A}_2 = \mathfrak{B}(H)$ then the composite system $\mathfrak{A} = \mathfrak{A}_1 \otimes \mathfrak{A}_2$ is called *hybrid system* (e.g. a quantum particle interacting with an experimental apparatus assumed to be classical). Considering $\mathscr{C}(X)$ as an algebra of functions (instead of an equivalent operator algebra), the tensor product $\mathscr{C}(X) \otimes \mathscr{B}(H)$ is the algebra of operator-valued functions on X :

$$\mathscr{C}(X) \otimes \mathscr{B}(H) = \{X \ni x \mapsto A(x) \in \mathscr{B}(H)\} \tag{3.13}$$

hence effects are operator-valued functions $x \mapsto A(x)$ such that $0 \leq A(x) \leq \mathbb{I}$ for any $x \in X$ and states are functional-valued functions $X \ni x \mapsto \rho(x) \in \mathscr{B}^*(H)$ s.t. $\rho(x) \geq 0$ and $\rho(x)(\mathbb{I}) = 1$ for any $x \in X$. On the other hand (3.13) can be identified as an operator algebra as well: Let K be a finite-dimensional Hilbert space with $\dim K = |X|$ and let $\{e_x\}_{x \in X}$ be an orthonormal basis of K. By means of the following definition:

$$\hat{A} := \sum_{x \in X} P_x \otimes A(x) \in \mathscr{B}(K \otimes H) \quad , \quad P_x := |e_x\rangle\langle e_x| \tag{3.14}$$

(3.13) can be identified as a subalgebra of $\mathscr{B}(K \otimes H)$.

In conclusion of this list of preliminary notions, let us recall the following definition:

Definition 3.2 The **partial trace** of $\sigma \in \mathscr{B}(H_1 \otimes H_2)$ w.r.t. H_2 (or H_1) is the unique operator $\text{tr}_{H_2}(\sigma) \in \mathscr{B}(H_1)$ such that:

$$\text{tr}[\text{tr}_{H_2}(\sigma)A] = \text{tr}[\sigma(A \otimes \mathbb{I}_2)] \qquad \forall A \in \mathscr{B}(H_1),$$

where trace on the left is calculated on H_1, trace on the right is calculated on $H_1 \otimes H_2$ and $\mathbb{I}_2$ is the identity operator on H_2.

If σ is a quantum state (identified to the corresponding density matrix) of the composite quantum system described in $\mathscr{B}(H_1 \otimes H_2)$, $\text{tr}_{H_2}(\sigma)$ is the *reduced density matrix* of the subsystem described in $\mathscr{B}(H_1)$, it represents the information that can be extracted measuring subsystem 1 and ignoring the other one. In terms of general observable algebras, if $\rho \in \mathfrak{S}(\mathfrak{A}_1 \otimes \mathfrak{A}_2)$ then the functional $\rho_1 \in \mathfrak{A}_1^*$, defined by $\rho_1(A) := \rho(A \otimes \mathbb{I})$ for all $A \in \mathfrak{A}_1$, is the *reduced state* of the system described in $\mathfrak{A}_1$.

3.2 Entanglement

3.2.1 Quantum Correlations and EPR Paradox

In the first section we have seen how the observable algebra of a composite system can be identified with an algebra of operators in a tensor product Hilbert space. In particular we are interested in n-level quantum systems, classical systems with a finite set of elementary events and hybrid systems that are relevant in quantum information theory.

Consider a bipartite system described in $\mathfrak{A} = \mathfrak{A}_1 \otimes \mathfrak{A}_2$ such that one of the subsystem is classical and the other one is quantum or classical, i.e. $\mathfrak{A}_1 = \mathscr{C}(X)$ and $\mathfrak{A}_2 = \mathscr{C}(Y), \mathfrak{B}(\mathsf{H})$. In other words the considered composite system is classical or hybrid.

Proposition 3.1 *Let* $\mathfrak{A} = \mathfrak{A}_1 \otimes \mathfrak{A}_2$ *be the observable algebra of a classical or hybrid composite system. Any state* $\rho \in \mathfrak{S}(\mathfrak{A})$ *can be written in the following form:*

$$\rho = \sum_{x \in X} \lambda_x \rho_x^{(1)} \otimes \rho_x^{(2)}, \tag{3.15}$$

where $\lambda_x \in [0, 1]$ *with* $\sum_x \lambda_x = 1$, $\rho_x^{(1)} \in \mathfrak{S}(\mathfrak{A}_1)$ *and* $\rho_x^{(2)} \in \mathfrak{S}(\mathfrak{A}_2)$.

Proof Since the considered system is classical or hybrid, let us assume $\mathfrak{A}_1 = \mathscr{C}(X)$. We know that any element f of $\mathscr{C}(X)$ can be identified with an operator in $\mathsf{H} \simeq \mathbb{C}^{|X|}$, we denote the corresponding operator with the same symbol $f :=$ $\sum_{x \in X} f(x)|e_x\rangle\langle e_x|$ where $\{e_x\}_{x \in X}$ is an orthonormal basis of H. Given a state $\rho \in \mathfrak{S}(\mathfrak{A}_1 \otimes \mathfrak{A}_2)$, let us define the following pair of functionals for any $x \in X$:

$$\rho_x^{(1)}(f) := f(x) \quad \text{and} \quad \rho_x^{(2)}(B) := \lambda_x^{-1}\rho(|e_x\rangle\langle e_x|\otimes B) \quad \text{with} \quad \lambda_x := \rho(|e_x\rangle\langle e_x|\otimes \mathbb{I}_K)$$

for all $f \in \mathscr{C}(X)$ and $B \in \mathfrak{A}_2 \subseteq \mathfrak{B}(\mathsf{K})$. Let us recall that $\mathfrak{A}_2$ can be a classical or a quantum algebra (in both cases it is an operator algebra on a Hilbert space K).

Proving $\rho_x^{(1,2)} \in \mathfrak{S}(\mathfrak{A}_{1,2})$ is straightforward. We can explicitly show that (3.15) holds:

$$\sum_{x \in X} \lambda_x \rho_x^{(1)}(f) \otimes \rho_x^{(2)}(B) = \sum_{x \in X} \lambda_x f(x) \lambda_x^{-1}\rho(|e_x\rangle\langle e_x| \otimes B)$$

$$= \sum_{x \in X} \rho(f(x)|e_x\rangle\langle e_x| \otimes B)$$

$$= \rho(f \otimes B),$$

for any $f \in \mathscr{C}(X)$ and $B \in \mathfrak{A}_2$.

Above result entails that a state of a (classical or hybrid) composite system can be always factorized in a statistical mixture of tensors in product form. This fact is not true if the composite system is completely quantum. In fact consider a pure state $\rho = |\Psi\rangle\langle\Psi| \in \mathfrak{S}(\mathfrak{B}(\mathsf{H}_1 \otimes \mathsf{H}_2))$, it does not admit a non-trivial convex decomposition, because it is an extreme element in the convex set of states, so ρ can be written in form (3.15) if and only if $\Psi = \psi_1 \otimes \psi_2$. Therefore a composite quantum system admits a class of physical states that are not of the form (3.15), i.e. these states are completely non-classical.

Let us denote the set of quantum states $\mathfrak{S}(\mathfrak{B}(H))$ simply by $\mathfrak{S}(H)$. We are in position to give a general definition:

Definition 3.3 Let $\rho \in \mathfrak{S}(H_1 \otimes H_2)$ be a state of a composite quantum system. ρ is said to be **separable** if it can be written in the form (3.15). Otherwise ρ is called **entangled**.

Let us observe that the correlations between physical systems in entangled states are apparently inconsistent with *Einstein's locality*: Consider the entangled pure state of a qubit pair identified by the vector

$$\Phi = \frac{1}{\sqrt{2}}(|0\rangle \otimes \psi + |1\rangle \otimes \varphi) \in H_1 \otimes H_2, \tag{3.16}$$

where $|0\rangle$ and $|1\rangle$ are orthonormal vectors in H_1 and $\psi, \varphi \in H_2$. If one performs the POVM-measurement $\{|0\rangle \langle 0|, |1\rangle \langle 1|\}$ on the first qubit (namely *local measurement*) then the state of the second qubit collapses[8] in the state $|\psi\rangle \langle \psi|$ or in the state $|\varphi\rangle \langle \varphi|$ according to the outcome of the local measurement. If we interpret the collapse as a consequence of an *action at distance* (producing a correlation between outcomes of measurement processes that can be spacelike separated) then we are violating the Einstein's locality principle (*EPR paradox* [10]). The experimental observation of these non-local correlations can be performed testing the violation of *Bell inequalities* [3], in this regard the first and most celebrated experiment was made by Aspect et al. in 1982 [2]. However this non-local feature cannot be used for communications faster than light (as proved in Sect. 3.4) then the principle of locality is preserved in terms of information transmission.

3.2.2 Sample of Separability Criteria

Understanding if a given quantum state is separable or entangled is not a simple issue in general and advanced techniques of tensor decomposition could be necessary in the *n*-partite case. Let us give a sample of some results to decide if the state of a bipartite quantum system is entangled or not.

Theorem 3.1 *For any entangled state* $\rho \in \mathfrak{S}(H_1 \otimes H_2)$ *there exists an operator* $A \in \mathfrak{B}(H_1 \otimes H_2)$ *such that* $\rho(A) < 0$ *and* $\widetilde{\rho}(A) \geq 0$ *for any separable state* $\widetilde{\rho} \in \mathfrak{S}(H_1 \otimes H_2)$.

Proof The claim is a consequence of Hahn-Banach theorem. The set $\mathfrak{S}^{sep}$ of separable states is closed and convex in $\mathfrak{B}^*(H_1 \otimes H_2)$ then for any state $\rho \notin \mathfrak{S}^{sep}$ there is an hyperplane which separates ρ and $\mathfrak{S}^{sep}$. More precisely, there exists a functional α on $\mathfrak{B}^*(H_1 \otimes H_2)$ such that $\alpha(\rho) < \beta \leq \alpha(\widetilde{\rho})$ for any $\widetilde{\rho} \in \mathfrak{S}^{sep}$, for

---

[8]In the sense of post-measurement state (3.4).

a constant β. Action of α can be represented by $\alpha(\rho) = \mathrm{tr}(A\rho) = \rho(A)$ for some $A \in \mathfrak{B}(H_1 \otimes H_2)$ (identifying $\mathfrak{B}^*(H_1 \otimes H_2)$ with $\mathfrak{B}(H_1 \otimes H_2)$). If $\beta = 0$ the claim is true, otherwise one has to consider the new functional $\alpha'(\rho) := \alpha(\rho) - \beta$.

The operator A is self-adjoint and it is called *entanglement witness*.

A necessary and sufficient condition for separability is given in terms of positive maps by the following theorem that is proved in [13].

Theorem 3.2 *A state $\rho \in \mathfrak{S}(H_1 \otimes H_2)$ is separable if and only if* $(\mathrm{id} \otimes F)\rho \geq 0$ *for any positive linear map $F : \mathfrak{B}^*(H_2) \to \mathfrak{B}^*(H_2)$ where* id *denotes the identity map on $\mathfrak{B}^*(H_1)$.*

Another separability criterion is based on the notion of partial transpose of a state in $\mathfrak{B}^*(H_1 \otimes H_2)$. Let A^T be the transpose of $A \in \mathfrak{B}(H_2)$ w.r.t. a fixed basis of H_2. We define the map $\varXi : \mathfrak{B}^*(H_2) \to \mathfrak{B}^*(H_2)$

$$(\varXi\rho)(A) := \rho(A^T). \tag{3.17}$$

The *partial transpose* in H_2 is defined as the map

$$\mathrm{id} \otimes \varXi : \mathfrak{B}^*(H_1 \otimes H_2) \to \mathfrak{B}^*(H_1 \otimes H_2).$$

Theorem 3.3 *Let* $\mathrm{id} \otimes \varXi : \mathfrak{B}^*(H_1 \otimes H_2) \to \mathfrak{B}^*(H_1 \otimes H_2)$ *be the partial transpose and assume* $\dim H_1 = 2$ *and* $\dim H_2 = 2, 3$. *A state $\rho \in \mathfrak{S}(H_1 \otimes H_2)$ is separable if and only if:*

$$(\mathrm{id} \otimes \varXi)\rho \geq 0. \tag{3.18}$$

For a proof of this statement see [20].

An important issue is quantifying entanglement, i.e. defining a good *entanglement measure* on quantum states. In this work we do not discuss the topic of entanglement measures, in this respect there is a huge literature, a survey on entanglement measures can be found in [21] for instance and a definition of an entanglement measure in terms of geometric formulation of QM is given in [18].

3.3 Quantum Channels

3.3.1 Completely Positive Maps

From an operational viewpoint, a *channel* is any process converting a physical system (input system with observable algebra $\mathfrak{A}_{in}$) into another one (output system with observable algebra $\mathfrak{A}_{out}$). If both involved systems are quantum then we talk about *quantum channels*, for instance free quantum evolution is a quantum channel where $\mathfrak{A}_{in} = \mathfrak{A}_{out} = \mathfrak{B}(H)$. A channel may convert a quantum system into a classical system, for example a measurement process is a channel where $\mathfrak{A}_{in} = \mathfrak{B}(\mathsf{H})$ and $\mathfrak{A}_{out} = \mathscr{C}(X)$, the classical system represents the experimental apparatus.

In order to give a general mathematical definition of channels we require that a channel can be described by a map transforming states of the input system into states of the output system:

$$T^* : \mathfrak{S}(\mathfrak{A}_{in}) \to \mathfrak{S}(\mathfrak{A}_{out}), \tag{3.19}$$

or equivalently by a map transforming effects of the output system into effects of the input system:

$$T : \mathfrak{E}(\mathfrak{A}_{out}) \to \mathfrak{E}(\mathfrak{A}_{in}), \tag{3.20}$$

so that $(T^*\rho)(E) = \rho(TE)$ for all $\rho \in \mathfrak{S}(\mathfrak{A}_{in})$ and $E \in \mathfrak{E}(\mathfrak{A}_{out})$. In other words we assume that a measurement process, given by a complete collection $\{E_i\}_i$ of effects, performed on the output system in the state $T^*\rho$ is equivalent (i.e. it provides exactly the same statistic) to the measurement process $\{TE_i\}_i$ on the input system in the state ρ. We say the map (3.19) describes the channel in the *Schrödinger picture* and the map (3.20) describes the channel in the *Heisenberg picture*.

Since $T^*\rho$ is a linear functional on $\mathfrak{A}_{out}$ then for any $\rho \in \mathfrak{S}(\mathfrak{A}_{in})$ and for all $A, B \in \mathfrak{E}(\mathfrak{A}_{out})$ we have:

$$\begin{aligned} \rho[T(aA + bB)] &= T^*\rho(aA + bB) \\ &= aT^*\rho(A) + bT^*\rho(B) \\ &= a\rho(TA) + b\rho(TB) \\ &= \rho(aTA + bTB), \end{aligned}$$

where $a \in [0, 1]$ and $b = 1 - a$. Let W, Z be elements of $\mathfrak{A}_{in}$, $W = Z$ if and only if $\rho(W) = \rho(Z)$ for every $\rho \in \mathfrak{S}(\mathfrak{A}_{in})$ (this fact holds for any C^*-algebra [16]), then $T(aA + bB) = aT(A) + bT(B)$ for any convex combination $aA + bB$ of effects, namely T is a convex-linear map on the convex set $\mathfrak{E}(\mathfrak{A}_{out})$ so it can be extended to a linear map on $\mathfrak{A}_{out}$. Then a channel is identified with a linear map $T : \mathfrak{A}_{out} \to \mathfrak{A}_{in}$ that is positive (i.e. $T(A) \geq 0$ for any $A \geq 0$) and unital (i.e. $T(\mathbb{I}_{out}) = \mathbb{I}_{in}$) so that T maps effects into effects and the dual map $T^* : \mathfrak{A}_{in}^* \to \mathfrak{A}_{out}^*$, defined by $T^*\rho(A) = \rho(TA)$, maps states into states.

We need another requirement on T in order to combine some channels in parallel. Consider two positive unital linear maps $T : \mathfrak{A}_{out} \to \mathfrak{A}_{in}$ and $S : \mathfrak{A}'_{out} \to \mathfrak{A}'_{in}$, we need to interpret the map:

$$(T \otimes S) : \mathfrak{A}_{out} \otimes \mathfrak{A}'_{out} \to \mathfrak{A}_{in} \otimes \mathfrak{A}'_{in}, \tag{3.21}$$

as a new channel processing composite systems. Therefore we must require $T \otimes S$ is positive as well, so we invoke the notion of *complete positivity*.

Definition 3.4 Let $\mathfrak{A}$ and $\mathfrak{D}$ be operator algebras. The linear map $T : \mathfrak{A} \rightarrow \mathfrak{D}$ is said to be **completely positive (CP-map)** if:

$$T \otimes \mathrm{id} : \mathfrak{A} \otimes \mathfrak{B}(\mathbb{C}^n) \rightarrow \mathfrak{D} \otimes \mathfrak{B}(\mathbb{C}^n), \qquad (3.22)$$

is positive for all $n \in \mathbb{N}$.

Complete positivity is a stronger property than positivity. However we have the following result involving classical algebras [22], let us recall that in our context a classical algebra is a C^*-algebra of functions on a finite set X which can be identified with an abelian operator algebra by means of (3.9):

Proposition 3.2 *Let $T : \mathfrak{A} \rightarrow \mathfrak{D}$ be a linear map and at least one of $\mathfrak{A}$ and $\mathfrak{D}$ be a classical algebra. Therefore T is CP if and only if T is positive.*

Hence complete positivity is not necessary to define *classical channels* that are just positive unital linear maps between classical algebras. Otherwise if both involved algebras are quantum then positivity does not imply complete positivity in general.

Example 3.1 Let $T : \mathfrak{B}(\mathbb{C}^2) \rightarrow \mathfrak{B}(\mathbb{C}^2)$ be the transposition of 2×2 complex matrices. It is obviously positive. Let us consider the action of $T \otimes \mathrm{id} : \mathfrak{B}(\mathbb{C}^2) \otimes \mathfrak{B}(\mathbb{C}^2) \rightarrow \mathfrak{B}(\mathbb{C}^2) \otimes \mathfrak{B}(\mathbb{C}^2)$ on the positive element:

$$A = \begin{pmatrix} 1 & 0 & 0 & 1 \\ 0 & 0 & 0 & 0 \\ 0 & 0 & 0 & 0 \\ 1 & 0 & 0 & 1 \end{pmatrix} \in \mathfrak{B}(\mathbb{C}^2) \otimes \mathfrak{B}(\mathbb{C}^2)$$

that yields a non-positive element:

$$(T \otimes \mathrm{id})(A) = \begin{pmatrix} 1 & 0 & 0 & 0 \\ 0 & 0 & 1 & 0 \\ 0 & 1 & 0 & 0 \\ 0 & 0 & 0 & 1 \end{pmatrix}$$

So T is positive but not CP.

We are in position to give the definitive definition of quantum channel:

Definition 3.5 Let $\mathfrak{A}_{out}$ and $\mathfrak{A}_{in}$ be quantum algebras. A **quantum channel** converting the input system described in $\mathfrak{A}_{in}$ into the output system described in $\mathfrak{A}_{out}$ is a unital CP-map $T : \mathfrak{A}_{out} \rightarrow \mathfrak{A}_{in}$.

Let us stress that the direction of mapping arrow is reversed w.r.t. the ordering of the process, on the other hand the ordering is preserved within the equivalent dual picture where quantum channel is represented by $T^* : \mathfrak{A}_{in}^* \to \mathfrak{A}_{out}^*$.

Some authors use the term *quantum operations* for unital CP maps of Definition 3.5. However a quantum channel is often distinguished from a general quantum operation since the former is trace preserving in Schrödinger picture and the latter is assumed to be simply trace non-increasing.

3.3.2 Stinespring Representation

Definition 3.4 can be directly re-formulated where $\mathfrak{A}$ and $\mathfrak{D}$ are abstract C^*-algebras instead of concrete operator algebras. In this spirit we state Stinespring dilation theorem for CP maps in a rather general version where $\mathfrak{A}$ is an abstract algebra and $\mathfrak{D}$ is the algebra of bounded operators in a Hilbert space.

Let us recall some notions: A C^*-algebra is said to be *unital* if it admits a unit $\mathbb{I}$ and $\| \mathbb{I} \| = 1$. Let $\mathfrak{A}$ and $\mathfrak{D}$ be C^*-algebras, an algebraic homomorphism $f : \mathfrak{A} \to \mathfrak{D}$ is called *$*$-homomorphism* if it preserves the involution: $f(A^{\dagger_\mathfrak{A}}) = f(A)^{\dagger_\mathfrak{D}}$ for any $A \in \mathfrak{A}$, where $\dagger_\mathfrak{A}$ and $\dagger_\mathfrak{D}$ are the involutions on $\mathfrak{A}$ and $\mathfrak{D}$ respectively.

Theorem 3.4 *Let $\mathfrak{A}$ be a unital C^*-algebra and H be a Hilbert space (also infinite dimensional). If $T : \mathfrak{A} \to \mathfrak{B}(\mathsf{H})$ is a CP map then there exist a Hilbert space K, a unital $*$-homomorphism $\pi : \mathfrak{A} \to \mathfrak{B}(\mathsf{K})$ and a bounded operator $V : \mathsf{H} \to \mathsf{K}$ satisfying $\| V \|^2 = \| \pi(1) \|$ such that:*

$$T(A) = V^\dagger \pi(A) V \qquad \forall A \in \mathfrak{A}. \tag{3.23}$$

Proof Let $[,] : \mathfrak{A} \otimes \mathsf{H} \times \mathfrak{A} \otimes \mathsf{H} \to \mathbb{C}$ be a sesquilinear form on $\mathfrak{A} \otimes \mathsf{H}$ defined by:

$$[A \otimes x, B \otimes y] := \langle T(B^\dagger A)x | y \rangle_\mathsf{H}, \qquad A, B \in \mathfrak{A}, \, x, y \in \mathsf{H}. \tag{3.24}$$

Let be $A_1, \ldots, A_n \in \mathfrak{A}$ and $x_1, \ldots, x_n \in \mathsf{H}$, so:

$$\left[\sum_{i=1}^n A_i \otimes x_i , \sum_{j=1}^n A_j \otimes x_j \right] = \sum_{i,j=1}^n \langle T(A_i^\dagger A_j) x_j | x_i \rangle \geq 0 \qquad \forall n \geq 1, \tag{3.25}$$

because T is a CP map. So the form $[,]$ is positive semidefinite and $N := \{u \in \mathfrak{A} \otimes \mathsf{H} : [u, u] = 0\}$ is a subspace of $\mathfrak{A} \otimes \mathsf{H}$. Let us define an inner product on $\frac{\mathfrak{A} \otimes \mathsf{H}}{N}$:

$$\langle [u] | [v] \rangle_\mathsf{K} := [u, v] \qquad [u], [v] \in \frac{\mathfrak{A} \otimes \mathsf{H}}{N}. \tag{3.26}$$

Let K be the Hilbert space defined as the completion of $\frac{\mathfrak{A} \otimes \mathsf{H}}{N}$.

For any $A \in \mathfrak{A}$ let us define the linear map $\pi(A) : \mathfrak{A} \otimes \mathsf{H} \to \mathfrak{A} \otimes \mathsf{H}$:

$$\pi(A) \left(\sum_{i=1}^{n} A_i \otimes x_i \right) := \sum_{i=1}^{n} A A_i \otimes x_i. \tag{3.27}$$

Following relation is a consequence of complete positivity of T and inequality $A^\dagger B^\dagger B A \leq \parallel B \parallel^2 A^\dagger A$ (that is true for any C^*-algebra):

$$[\pi(A)u, \pi(A)u] \leq \parallel A \parallel^2 [u, u] \qquad \forall u \in \mathfrak{A} \otimes \mathsf{H}. \tag{3.28}$$

Then N belongs to $Ker(\pi(A))$ and $\pi(A)$ defines a bounded operator in $\frac{\mathfrak{A} \otimes \mathsf{H}}{N}$ therefore in K. One can check $\pi : \mathfrak{A} \to \mathfrak{B}(\mathsf{K})$ is a unital $*$-homomorphism by direct inspection.

Let us define the operator $V : \mathsf{H} \to \mathsf{K}$ by:

$$V x := [\mathbb{I} \otimes x] \qquad \forall x \in \mathsf{H}, \tag{3.29}$$

where $\mathbb{I}$ denotes the unit of $\mathfrak{A}$.

$$\parallel V x \parallel^2 = [\mathbb{I} \otimes x, \mathbb{I} \otimes x] = \langle T(\mathbb{I}) x | x \rangle_\mathsf{H} = \langle \sqrt{T(\mathbb{I})} x | \sqrt{T(\mathbb{I})} x \rangle_\mathsf{H} = \parallel \sqrt{T(\mathbb{I})} x \parallel^2 \qquad \forall x \in \mathsf{H},$$

thus $\parallel V \parallel^2 = \parallel T(\mathbb{I}) \parallel$. We are in position to prove the claim (3.23):

$$\langle V^\dagger \pi(A) V x | y \rangle_\mathsf{H} = \langle \pi(A)[\mathbb{I} \otimes x] | [\mathbb{I} \otimes y] \rangle_\mathsf{K} = \langle [A \otimes x] | [\mathbb{I} \otimes y] \rangle_\mathsf{K} = [A \otimes x, \mathbb{I} \otimes y] = \langle T(A) x | y \rangle_\mathsf{H},$$

for any $x, y \in \mathsf{H}$ and for all $A \in \mathfrak{A}$.

If $\pi : \mathfrak{A} \to \mathfrak{B}(\mathsf{K})$ is a unital $*$-homomorphism and $V : \mathsf{H} \to \mathsf{K}$ is a bounded operator then $T(A) = V^\dagger \pi(A) V$ defines a CP-map from $\mathfrak{A}$ to $\mathfrak{B}(\mathsf{H})$. Therefore the Stinespring theorem completely characterizes CP-maps.

Let (K, π, V) be a Stinespring triple of the CP-map T. If

$$\pi(\mathfrak{A}) V \mathsf{H} = \{ \pi(A) V x : A \in \mathfrak{A}, x \in \mathsf{H} \}$$

has dense span in K then (K, π, V) is called *minimal Stinespring representation* of T. This decomposition is unique up to a unitary equivalence.

We are interested in a particular case of Stinespring statement: Applying the constructive proof of Theorem 3.4 we can prove the following result:

Theorem 3.5 *Let H_1 and H_2 be finite-dimensional Hilbert spaces. For any CP-map $T : \mathfrak{B}(\mathsf{H}_1) \to \mathfrak{B}(\mathsf{H}_2)$ there are a finite-dimensional Hilbert space K and an operator $V : \mathsf{H}_2 \to \mathsf{H}_1 \otimes \mathsf{K}$ such that:*

$$T(A) = V^\dagger (A \otimes \mathbb{I}_\mathsf{K}) V \qquad \forall A \in \mathfrak{B}(\mathsf{H}_1). \tag{3.30}$$

Let us choose an orthonormal basis $\{\Psi_i\}_i$ of K, the *Kraus operators* $V_\alpha : \mathsf{H}_2 \to \mathsf{H}_1$ are defined by the following relation:

$$\langle \psi | V_\alpha \varphi \rangle_{\mathsf{H}_1} = \langle \psi \otimes \Psi_\alpha | V \varphi \rangle_{\mathsf{H}_1 \otimes \mathsf{K}} \qquad \psi \in \mathsf{H}_1 , \; \varphi \in \mathsf{H}_2, \qquad (3.31)$$

then any quantum channel admits a *Kraus representation*:

$$T(A) = \sum_{\alpha=1}^{n} V_\alpha^\dagger A V_\alpha \qquad n \leq \dim(\mathsf{H}_1) \dim(\mathsf{H}_2). \qquad (3.32)$$

Proposition 3.3 *Let $T : \mathfrak{A} \to \mathfrak{A}$ be a quantum channel and $\{V_\alpha\}_\alpha$ be a Kraus representation of T. $T(A) = A$ if and only if $[A, V_\alpha] = 0$ for all α.*

Proof If $[A, V_\alpha] = 0$ $\forall \alpha$ then $T(A) = \sum_\alpha V_\alpha^\dagger A V_\alpha = \sum_\alpha V_\alpha^\dagger V_\alpha A = A$ because $\sum_\alpha V_\alpha^\dagger V_\alpha = \mathbb{I}$ (i.e. T is unital). Conversely, if $T(A) = A$ then we have:

$$T(A)^\dagger T(A) - A^\dagger T(A) - T(A)^\dagger A + A^\dagger A = 0.$$

By the Kraus representation (3.32) of T we obtain:

$$\sum_\alpha [A, V_\alpha]^\dagger [A, V_\alpha] = 0,$$

since all terms in the sum are non-negative, $[A, V_\alpha] = 0, \forall \alpha$.

This result implies that the Kraus operators of T belong to the commutant of the subalgebra $\mathfrak{A}_T \subseteq \mathfrak{A}$ of invariant elements w.r.t. T. We will use this fact further to prove a general version of no-cloning theorem.

3.3.3 Noisy Channels

Transmission of quantum information (encoded in the state of a photon for instance) on long distances can be described by a channel $T : \mathfrak{B}(\mathsf{H}) \to \mathfrak{B}(\mathsf{H})$, so the received information is represented by the state $T^* \rho$ when sent information is the state ρ. If the channel T is *ideal* then there are not information losses, an example of ideal channel is the identity map $T = \mathrm{id}_{\mathfrak{B}(\mathsf{H})}$, more generally ideal channels correspond to invertible channels (i.e. a state can be altered but the initial information can be completely restored). If $T : \mathfrak{B}(\mathsf{H}) \to \mathfrak{B}(\mathsf{H})$ is invertible then $T^* \rho = U \rho U^\dagger$ for some unitary operator U on H by the Stinespring representation of T. In other words the channel acts as a free evolution, i.e. the system (e.g. a photon travelling in a waveguide) is isolated and does not interact with the environment.

On the other hand there are *noisy channels* describing an information processing where the considered quantum system interacts with the environment. The general structure of a noisy channel $T : \mathfrak{B}(H) \to \mathfrak{B}(H)$ in Schrödinger picture is:

$$T^*\rho = \mathrm{tr}_K(U\rho \otimes \rho_0 U^\dagger), \tag{3.33}$$

where K is the Hilbert space of the environment, U is a unitary operator in $H \otimes K$ and ρ_0 is the initial state of the environment. So we are assuming that the action of a noisy channel is a reduced dynamics of the composite system *System+Environment* as an isolated system. This assumption is justified by the following result which entails that the action of any (noisy) channel can be modeled by an interaction with another system prepared in a pure state, called *Ancilla system*.

Theorem 3.6 *If* $T : \mathfrak{B}(H) \to \mathfrak{B}(H)$ *is a quantum channel then there exist a Hilbert space* K, *a pure state* $\rho_0 \in \mathfrak{S}(K)$ *and a unitary operator* $U : H \otimes K \to H \otimes K$ *such that:*

$$T^*\rho = \mathrm{tr}_K(U\rho \otimes \rho_0 U^\dagger). \tag{3.34}$$

Proof Consider a Stinespring representation of T:

$$T(A) = V^\dagger(A \otimes \mathbb{I}_K)V \qquad \forall A \in \mathfrak{B}(H),$$

where $V : H \to H \otimes K$. Let be $\psi \in K$ and consider an operator U in $H \otimes K$ such that $U(\phi \otimes \psi) = V\phi$ for any $\phi \in H$. Since T is unital then V is an isometry, in fact $T(\mathbb{I}_H) = V^\dagger(\mathbb{I}_H \otimes \mathbb{I}_K)V = V^\dagger V = \mathbb{I}_H$, so it can be always extended to a unitary $U : H \otimes K \to H \otimes K$. Let $\{e_i\}_i$ be an orthonormal basis of H and $\{f_j\}_j$ be an orthonormal basis of K:

$$(T^*\rho)(A) = \mathrm{tr}(T(A)\rho) = \mathrm{tr}(V^\dagger(A \otimes \mathbb{I}_K)V\rho) = \sum_i \langle V\rho e_i|(A \otimes \mathbb{I}_K)Ve_i\rangle =$$

$$= \sum_{i,j}\langle U(\rho \otimes |\psi\rangle\langle\psi|)(e_i \otimes f_j)|(A \otimes \mathbb{I}_K)U(e_i \otimes f_j)\rangle = \mathrm{tr}\left[\mathrm{tr}_K(U(\rho \otimes |\psi\rangle\langle\psi|)U^\dagger)A\right].$$

The claim is proved for $\rho_0 = |\psi\rangle\langle\psi|$.

3.4 Quantum Communications

Now we are in position to use quantum channels as a model for processing and transmitting information. In this section we give a short review about *impossible machines* like the *quantum cloner*. Then we will present two celebrated entanglement-assisted processes for information transmission: *Quantum teleportation* and *dense coding*.

3.4.1 Information Processing

Roughly speaking, classical information is encoded in the physical state of a classical system, e.g. a bit whose state is 0 or 1. On the other hand quantum information is encoded in the physical state of a quantum system, e.g. a qubit whose state is a linear combination of two orthonormal states $|0\rangle$ and $|1\rangle$ (vertical and horizontal polarization of a photon for instance).

Let us focus on the relationship between classical and quantum channels in order to convert quantum systems (quantum information) into classical systems (classical information) and viceversa. We basically follow the approach adopted in [14] for instance. Let us recall that the observable algebra $\mathscr{C}(X)$ of a classical system is given by the complex functions on X, where X is the finite set of elementary events. $\mathscr{C}(X)$ can be identified with an operator algebra, in particular an abelian subalgebra of $\mathfrak{B}(\mathsf{H})$, with $\dim \mathsf{H} = |X|$, in view of (3.9). We denote the classical operator algebra with the same symbol $\mathscr{C}(X)$, once fixed an orthonormal basis $\{e_x\}_{x \in X}$ of H, orthogonal projectors $\{P_x = |e_x\rangle\langle e_x|\}_{x \in X}$ form a basis of the operator algebra $\mathscr{C}(X)$.

A classical channel is a linear map $T : \mathscr{C}(X) \to \mathscr{C}(Y)$ that is positive and unital. Identifying $\mathscr{C}(X)$ and $\mathscr{C}(Y)$ as subalgebras of $\mathfrak{B}(\mathsf{H})$ and $\mathfrak{B}(\mathsf{K})$ with orthonormal basis $\{P_x\}_{x \in X}$ and $\{Q_y\}_{y \in Y}$ respectively, T is completely characterized by its matrix elements:

$$T_{xy} := (Q_y | T(P_x))_{\mathsf{K}}, \tag{3.35}$$

where $(A|B)_{\mathsf{K}} := \mathrm{tr}(A^* B)$ for $A, B \in \mathfrak{B}(\mathsf{K})$. Since:

$$\sum_{x \in X} T_{xy} = \left(Q_y \left| T \left(\sum_{x \in X} P_x \right) \right. \right)_{\mathsf{K}} = (Q_y | T(\mathbb{I}_{\mathsf{H}}))_{\mathsf{K}} = (Q_y | \mathbb{I}_{\mathsf{K}})_{\mathsf{K}} = \mathrm{tr}(Q_y) = 1,$$

$x \mapsto T_{xy}$ is a probability distribution on X. Hence T_{xy} is interpreted as the probability to get the output $x \in X$ when the input is $y \in Y$. Therefore there are no errors in transmission if and only if $\mathscr{C}(X) = \mathscr{C}(Y)$ and $T_{xy} = \delta_{xy}$. So an ideal classical channel is described by $\mathrm{id}_{\mathscr{C}(X)}$.

Now let us consider a channel $E : \mathfrak{A}_{out} \to \mathfrak{A}_{in}$, where $\mathfrak{A}_{out} = \mathscr{C}(X) \subset \mathfrak{B}(\mathsf{H})$ and $\mathfrak{A}_{in} = \mathfrak{B}(\mathsf{K})$, converting a quantum system into a classical system. Let $\{P_x\}_{x \in X}$ be a basis of $\mathscr{C}(X)$ as usual and $E_x := E(P_x)$ for any $x \in X$. The family $\{E_x\}_{x \in X}$ is made by positive operators in $\mathfrak{B}(\mathsf{K})$ such that:

$$\sum_{x \in X} E_x = E \left(\sum_{x \in X} P_x \right) = E(\mathbb{I}_{\mathsf{H}}) = \mathbb{I}_{\mathsf{K}},$$

so $\{E_x\}_{x \in X}$ is a POVM on the Hilbert space K. Therefore a general quantum/classical information converter can be always interpreted as a measurement process. In other words the only way to convert quantum information into classical information is performing measurements on a quantum system. In the Schrödinger picture the channel $E^* : \mathfrak{B}^*(\mathsf{H}) \to \mathscr{C}^*(X)$ converts the quantum state ρ into the classical state $E^*\rho = p$ that is a probability distribution on X given by:

$$p(x) = \mathrm{tr}(E_x \rho). \qquad (3.36)$$

A classical/quantum information converter is a channel $F^* : \mathscr{C}^*(X) \to \mathfrak{B}^*(\mathsf{H})$. Let $\delta_x \in \mathscr{C}^*(X)$ be the Dirac measure centered in $x \in X$, so the channel F^* is completely defined by the family of quantum states $\{F^*(\delta_x)\}_{x \in X} \equiv \{\rho_x\}_{x \in X}$. Then the most general way to convert classical information into quantum information is a *parameter-dependent preparation* $X \ni x \mapsto \rho_x \in \mathfrak{S}(\mathsf{H})$. The simplest example is mapping a bit into a qubit: $0 \mapsto \rho_0 = |0\rangle \langle 0|$ and $1 \mapsto \rho_1 = |1\rangle \langle 1|$, where $\{|0\rangle, |1\rangle\}$ is an orthonormal basis of $\mathbb{C}^2$.

A channel $L^* : \mathfrak{B}^*(\mathsf{H}) \otimes \mathscr{C}^*(X) \to \mathfrak{B}^*(\mathsf{K})$ converting a hybrid system into a quantum system describes a quantum operation whose action varies according to classical information. For this reason we call it *parameter-dependent operation*. Defining $L_x^* : \mathfrak{B}^*(\mathsf{H}) \to \mathfrak{B}^*(\mathsf{K})$:

$$L_x^* \rho := L^*(\rho \otimes \delta_x), \qquad (3.37)$$

for any $x \in X$, the action of the channel is $L^*(\rho \otimes p) = \sum_{x \in X} p(x) L_x^* \rho$ then it is represented by a one-parameter family of quantum channels $\{L_x^*\}_{x \in X}$.

3.4.2 Relevant No-Go Theorems: Impossible Machines

In this section we have a short look to some no-go theorems which prevent the construction of some hypothetical devices called *impossible machines*. For instance one can suppose to exploit quantum correlations to attempt superluminal communications. In fact if two quantum systems A and B are entangled then an operation performed on A perturbs the state of B regardless of their spatial separation. Can this feature be applied to construct a device that is able to allow superluminal communications? The answer is *no*, quantum correlations cannot be exploited to communicate any information beyond the limit imposed by the speed of light. This fact is implied by the following result even if our whole discussion is non-relativistic.

Theorem 3.7 (No-communication) *Let $\mathfrak{A}$ and $\mathfrak{B}$ be the observable algebras of the quantum systems A and B and $\mathfrak{C}$ be the observable algebra of an arbitrary physical system C (classical, quantum, hybrid). Let $T : \mathfrak{C} \to \mathfrak{A}$ be a channel converting the* *system A into the system C. Then for any state $\rho \in \mathfrak{S}(\mathfrak{A} \otimes \mathfrak{B})$ following identity holds:*

$$\rho(\mathbb{I}_{\mathfrak{A}} \otimes B) = [T^* \otimes \mathrm{id}_{\mathfrak{B}^*}\rho](\mathbb{I}_{\mathfrak{C}} \otimes B) \qquad \forall B \in \mathfrak{B}, \qquad (3.38)$$

where $\mathbb{I}_{\mathfrak{A}}$ and $\mathbb{I}_{\mathfrak{C}}$ are the units in $\mathfrak{A}$ and $\mathfrak{C}$ respectively.

Proof A state $\rho \in \mathfrak{S}(\mathfrak{A} \otimes \mathfrak{B})$ is a functional in $(\mathfrak{A} \otimes \mathfrak{B})^*$ then it can be decomposed as a finite sum[9]:

$$\rho = \sum_i \alpha_i \otimes \beta_i \qquad \alpha_i \in \mathfrak{A}^*, \ \beta_i \in \mathfrak{B}^*.$$

The claim can be explicitly proved:

$$[T^* \otimes \mathrm{id}_{\mathfrak{B}^*}\rho](\mathbb{I}_{\mathfrak{C}} \otimes B) = \sum_i T^*\alpha_i(\mathbb{I}_{\mathfrak{C}}) \otimes \beta_i(B) = \sum_i \alpha_i(T(\mathbb{I}_{\mathfrak{C}})) \otimes \beta_i(B) = \rho(\mathbb{I}_{\mathfrak{A}} \otimes B),$$

for any $B \in \mathfrak{B}$.

This theorem has a strong physical interpretation: If ρ is the quantum state of the composite system $A + B$, then the functional $\rho_B : B \mapsto \rho(\mathbb{I}_{\mathfrak{A}} \otimes B)$ is the reduced state of the subsystem B. This state turns out to be indistinguishable from $\rho_B' : B \mapsto [(T^* \otimes \mathrm{id}_{\mathfrak{B}^*})\rho](\mathbb{I}_{\mathfrak{C}} \otimes B)$ that is the reduced state of B after that an arbitrary operation T has been performed on A (that can be a quantum operation like the coupling with another quantum system, or a controlled time evolution, or a projective measurement, or a destructive measurement, etc.). In other words no information can be sent to Bob by any Alice's operation on her subsystem.

Now let us focus on the following problem: *Can a quantum state be classically teleported?* More precisely, consider a quantum system prepared in the state $\rho \in \mathfrak{S}(H)$. Suppose Alice performs a measurement process on it that is described by the POVM $E = \{E_x\}_{x \in X} \subset \mathfrak{B}(H)$. Then she communicates the outcome $x \in X$ to Bob who applies a parameter-dependent preparation $x \mapsto \rho_x \in \mathfrak{S}(H)$ on his copy of the quantum system. Assuming this procedure is repeated many times the final state that Bob re-constructs is given by:

$$\rho_B = \sum_{x \in X} \mathrm{tr}(E_x\rho)\rho_x. \qquad (3.39)$$

A classical teleportation has been implemented if $\rho_B = \rho$. We ask if there exist a measurement process and a parameter-dependent preparation such that an arbitrary quantum state can be transmitted even if only a classical channel is available. Let us formulate the notion of classical teleportation in a more mathematical way:

---

[9]We assume that the considered observable algebras are operator algebras on finite-dimensional Hilbert spaces as usual.

Definition 3.6 Let $E : \mathscr{C}(X) \to \mathfrak{B}(\mathsf{H})$ be a quantum measurement, $T : \mathscr{C}(Y) \to \mathscr{C}(X)$ be a classical channel and $F : \mathfrak{B}(\mathsf{H}) \to \mathscr{C}(Y)$ be a parameter-dependent preparation. The pair (E, T, F) is called **classical teleportation scheme** if $ETF = id_{\mathfrak{B}(\mathsf{H})}$ (or $F^* T^* E^* = id_{\mathfrak{B}^*(\mathsf{H})}$).

In a classical teleportation scheme, the measurement process E maps a (generally mixed) quantum state ρ of a system S into a probability distribution p on the set of outcomes X and the parameter-dependent preparation F converts the classical probability distribution p into a quantum state of a *copy of S* which coincides with the initial state ρ:

$$\rho \mapsto E^*(\rho) = p \mapsto F^*(p) = \rho, \tag{3.40}$$

with $p(x) = \mathrm{tr}(E_x \rho)$ and $F^*(p) = \sum_{x \in X} p(x)\rho_x$, where E_x are the elements of the POVM E and $x \mapsto \rho_x$ is the parameter-dependent preparation as discussed in Sect. 3.4.1.

Let us stress that the systems S and the *copy of S* are not complex systems but rather identical particles (photons, electrons, atoms, molecules, quasiparticles etc.). Therefore by indistinguishability of identical particles, teleportation of a quantum state can be intended as the teleportation of the quantum system itself in the sense that any physical property of S is reproduced into *copy of S*.

Theorem 3.8 (No-teleportation) *No classical teleportation scheme exists.*

In order to prove the statement above, let us use the following argument: If the classical teleportation was possible then we could perfectly duplicate an arbitrary quantum state. In fact a classical teleportation scheme allows to convert whole quantum information encoded in a quantum state into classical information, then a classical copying procedure can be performed and finally two perfect copies of the initial quantum state can be obtained:

$$\textit{Classical teleportation} \implies \textit{Quantum cloning} \tag{3.41}$$

However a general cloning procedure of a quantum states is not possible [25]. Let us give a proof of the no-cloning theorem in a rather general scenario [15]: Let $\mathfrak{A}$ be a quantum observable algebra, a *cloning machine* is defined by a quantum channel $C^* : \mathfrak{A}^* \to \mathfrak{A}^* \otimes \mathfrak{A}^*$ such that for any state $\rho \in \mathfrak{S}(\mathfrak{A})$ we have

$$C^*\rho(A \otimes \mathbb{I}) = C^*\rho(\mathbb{I} \otimes A) = \rho(A) \qquad \forall A \in \mathfrak{A}. \tag{3.42}$$

The cloning is assumed to be physically realized by the coupling between the input system and an apparatus that is arbitrarily complex and whose dynamics is not described by the channel C so its observable algebra is not considered.

Theorem 3.9 (No-cloning) *No quantum cloning machine exists, i.e. there not exists a quantum channel satisfying property (3.42).*

Proof Consider an operator algebra $\mathfrak{C} \subset \mathfrak{B}(\mathsf{H})$, its commutant is defined as $\mathfrak{C}' :=$ $\{A \in \mathfrak{B}(\mathsf{H}) : [A, B] = 0, \forall B \in \mathfrak{C}\}$. Given a quantum channel $T : \mathfrak{A} \rightarrow \mathfrak{A}$ we consider the subalgebra where T acts as the identity $\mathfrak{A}_T := \{A \in \mathfrak{A} : T(A) = A\}$ and define the subset of invariant states $\mathfrak{S}_T(\mathfrak{A}) := \{\rho \in \mathfrak{S}(\mathfrak{A}) : T^*(\rho) = \rho\}$. Let us denote the set of density matrices in $\mathfrak{S}(\mathfrak{A})$ commuting with all elements of $\mathfrak{S}_T(\mathfrak{A})$ as $\mathfrak{S}'_T(\mathfrak{A})$.

Let us assume the existence of a cloning machine $C : \mathfrak{A} \otimes \mathfrak{A} \rightarrow \mathfrak{A}$ and define two other channels operating on $\mathfrak{A}$:

$$R(A) := C(A \otimes \mathbb{I}) \qquad T(A) := C(\mathbb{I} \otimes A) \qquad \forall A \in \mathfrak{A},$$

R can be interpreted as the action of the cloning machine on the input system and T as the cloning operation itself. Let us prove that $T(\mathfrak{A}) \subseteq \mathfrak{A}'_R$, where $\mathfrak{A}'_R$ is the commutant of $\mathfrak{A}_R := \{A \in \mathfrak{A} : R(A) = A\}$: Since $\mathfrak{A} = \mathfrak{B}(\mathsf{H})$, i.e. a quantum algebra, we consider the Kraus operators $V_\alpha : \mathsf{H} \rightarrow \mathsf{H} \otimes \mathsf{H}$ of C, for any α we have $V_\alpha A = (A \otimes \mathbb{I})V_\alpha$ for every $A \in \mathfrak{A}_R$ as a consequence of Proposition 3.3. Hence:

$$C(A \otimes B) = AC(\mathbb{I} \otimes B) = C(\mathbb{I} \otimes B)A \qquad \forall A \in \mathfrak{A}_R, \ \forall B \in \mathfrak{A}.$$

The perfect cloning can be required imposing that $\rho \in \mathfrak{S}_T(\mathfrak{A}) \cap \mathfrak{S}_R(\mathfrak{A})$ for every $\rho \in \mathfrak{S}(\mathfrak{A})$, i.e. $\mathfrak{S}_T(\mathfrak{A}) = \mathfrak{S}_R(\mathfrak{A}) = \mathfrak{S}(\mathfrak{A})$. Let us prove that the latter fact is impossible: Since $T(\mathfrak{A}) \subseteq \mathfrak{A}'_R$ then we have that $\mathfrak{A}_T \subseteq \mathfrak{A}'_R$ so the set of cloneable density matrices $\mathfrak{S}_T(\mathfrak{A}) \cap \mathfrak{S}_R(\mathfrak{A})$ belongs to $\mathfrak{S}'_R(\mathfrak{A}) \cap \mathfrak{S}_R(\mathfrak{A})$ that is a commutative set of density matrices which cannot correspond to $\mathfrak{S}(\mathfrak{A})$. Therefore no general cloning procedure exists.

No-cloning theorem prevents the arbitrary copying of quantum states, as a consequence a quantum state cannot be classically teleported by negation of (3.41). In particular quantum no-cloning has a remarkable impact on quantum communications, in fact it implies that an eavesdropper cannot gain information from unknown quantum states by means of a cloning procedure. Thus an eavesdropper must be content to get an approximate copy of a state (e.g. [7]), on the other hand if he performs measurement processes then he disturbs the quantum transmission, in fact a quantum measurement yields no information gain without disturbance (in the general sense described in [6] for instance). So we can formulate one of the paradigms of quantum cryptography: *A quantum transmission can be intercepted but no information can be extracted without detectable effects.*

3.4.3 Quantum Teleportation

Classical teleportation is not allowed by foundations of QM, however a teleportation scheme based on entanglement can be constructed. Exploiting entanglement as a resource a quantum state can be teleported. This section is focused on this mechanism.

Suppose the aim is the transmission of the quantum state $\rho \in \mathfrak{S}(\mathsf{H})$ of a *target system* from Alice to Bob. Suppose an ideal classical channel is allowed, i.e. it is described by the identity map on the observable algebra $\mathscr{C}(X)$ so a classical datum $x \in X$ can be transmitted without errors. Consider a quantum composite system formed by two identical systems that we call *ancillas* prepared in the entangled state $\sigma \in \mathfrak{S}(\mathsf{K} \otimes \mathsf{K})$, an ancilla is sent to Alice and the other to Bob.

Le $\{E_x\}_{x \in X}$ be a POVM in $\mathfrak{B}(\mathsf{H} \otimes \mathsf{K})$ which describes a measurement process on the composite system formed by the target system and Alice's ancilla. This POVM defines a channel $E : \mathscr{C}(X) \to \mathfrak{B}(\mathsf{H} \otimes \mathsf{K})$ as discussed in Sect. 3.4.1.

Let $L : \mathfrak{B}(\mathsf{H}) \to \mathscr{C}(X) \otimes \mathfrak{B}(\mathsf{K})$ be a parameter-dependent operation which describes an operation performed on Bob's ancilla depending on the outcome of the measurement E.

Definition 3.7 Let $T^* : \mathfrak{B}^*(\mathsf{H} \otimes \mathsf{K}) \otimes \mathfrak{B}^*(\mathsf{K}) \to \mathfrak{B}^*(\mathsf{H})$ be the quantum channel defined by

$$T^* := L^* \circ (E^* \otimes id_{\mathfrak{B}^*(\mathsf{K})}),$$

where E is a measurement process and L is a parameter-dependent operation. If there exists a state $\sigma \in \mathfrak{S}(\mathsf{K} \otimes \mathsf{K})$ such that $T^*(\rho \otimes \sigma) = \rho$ for any $\rho \in \mathfrak{S}(\mathsf{H})$ then the triple (E, L, σ) is called **quantum teleportation scheme**.

Let us show the existence of quantum teleportation by the concrete construction of a particular scheme. Consider the finite set $X = \{0, 1, 2, \ldots, n^2 - 1\}$ and the Hilbert spaces H and K, where target and ancilla are described, such that $\mathsf{H} \simeq \mathsf{K} \simeq \mathbb{C}^n$. Let $\sigma \in \mathfrak{S}(\mathsf{K} \otimes \mathsf{K})$ be a maximally entangled state[10] of Alice and Bob's ancillas:

$$\sigma := |\Psi\rangle\langle\Psi| \qquad \Psi := \frac{1}{\sqrt{n}} \sum_{i=1}^{n} \varphi_i \otimes \varphi_i \qquad (3.43)$$

where $\{\varphi_i\}_i$ is an orthonormal basis of $\mathbb{C}^n$.

Consider a one-parameter family of unitary operators $\{U_x\}_{x \in X}$ on $\mathbb{C}^n$ such that $\mathrm{tr}(U_x^\dagger U_y) = n\, \delta_{xy}$ and let $\{\Phi_x\}_{x=1,\ldots,n^2-1}$ be an orthonormal basis of $\mathsf{H} \otimes \mathsf{K}$ defined by $\Phi_x := (U_x \otimes \mathbb{I})\Psi$. We define the measurement process $E^* : \mathfrak{B}^*(\mathsf{H} \otimes \mathsf{K}) \to \mathscr{C}^*(X)$ by means of the POVM $\{|\Phi_x\rangle\langle\Phi_x|\}_{x \in X}$:

$$E^* : \rho \mapsto p \quad \text{where} \quad p(x) := \mathrm{tr}(|\Phi_x\rangle\langle\Phi_x|\rho). \qquad (3.44)$$

---

[10]If $\mathsf{K} \simeq \mathbb{C}^2$ and $\{|0\rangle, |1\rangle\}$ denotes an orthonormal basis of K then the maximally entangled states are the four *Bell states*: $\Phi^\pm = \frac{1}{\sqrt{2}}(|00\rangle \pm |11\rangle)$ and $\Psi^\pm \frac{1}{\sqrt{2}}(|01\rangle \pm |10\rangle)$ where $|i\rangle \otimes |j\rangle \equiv |ij\rangle$. The Bell states give an orthonormal basis of the Hilbert space of a qubit pair that is often convenient.

Furthermore we define a parameter-dependent operation $L^* : \mathscr{C}^*(X) \otimes \mathscr{B}^*(K) \rightarrow \mathscr{B}^*(H)$ by:

$$L^* : p \otimes \rho \mapsto \sum_{x=0}^{n^2-1} p(x) U_x \rho U_x^\dagger. \tag{3.45}$$

Let us verify that (E, L, σ) is a quantum teleportation scheme, i.e. $T^*\rho = L^*(E^* \otimes \mathrm{id})\rho = \rho$ for any $\rho \in \mathfrak{S}(H)$.

Proposition 3.4 *Consider the quantum measurement defined in (3.44), the parameter-dependent operation defined in (3.45) and the quantum state defined in (3.43) then*

$$T^*(\rho \otimes \sigma) = L^*(E^* \otimes \mathrm{id})(\rho \otimes \sigma) = \rho \qquad \forall \rho \in \mathscr{B}^*(H).$$

Proof The action of T^* is:

$$[T^*(\rho \otimes \sigma)](A) = \mathrm{tr}[T^*(\rho \otimes \sigma)A] = \mathrm{tr}[L^*(E^* \otimes \mathrm{id})(\rho \otimes \sigma)A] \qquad \forall A \in \mathscr{B}(H)$$

denoting the functional $T^*(\rho \otimes \sigma) \in \mathscr{B}^*(H)$ and its Riesz-representative operator with the same symbol. In particular:

$$(E^* \otimes \mathrm{id})(\rho \otimes \sigma) = \mathrm{tr}_{1,2}[(|\Phi_j\rangle\langle\Phi_j| \otimes \mathbb{I})(\rho \otimes \sigma)],$$

where $\mathrm{tr}_{1,2}$ denotes the trace operation over the first two tensor factors.[11] By definition of L^*:

$$\mathrm{tr}[T^*(\rho \otimes \sigma)A] = \mathrm{tr}\left[\sum_{x=0}^{n^2-1} U_x \mathrm{tr}_{1,2}[(|\Phi_x\rangle\langle\Phi_x| \otimes \mathbb{I})(\rho \otimes \sigma)] U_x^\dagger A \right] =$$

$$= \sum_{x=0}^{n^2-1} \mathrm{tr}\left[(\rho \otimes \sigma)(|\Phi_x\rangle\langle\Phi_x| \otimes U_x^\dagger A U_x) \right],$$

according to definition of partial trace. Requiring $\Phi_x = (U_x \otimes \mathbb{I})\Psi$ for any $x \in X$, let us prove the following fact [24]:

$$\sum_{x=0}^{n^2-1} \mathrm{tr}\left[(\rho \otimes \sigma)(|\Phi_x\rangle\langle\Phi_x| \otimes U_x^\dagger A U_x) \right] = \mathrm{tr}(\rho A). \tag{3.46}$$

---

[11] The partial trace over the first two tensor factors $\mathrm{tr}_{1,2} : \mathscr{B}(H_1 \otimes H_2 \otimes H_3 \otimes \cdots \otimes H_n) \rightarrow \mathscr{B}(H_3 \otimes \cdots \otimes H_n)$ is defined as $\mathrm{tr}_{1,2}(A_1 \otimes A_2 \otimes A_3 \otimes \cdots \otimes A_n) := \mathrm{tr}(A_1)\mathrm{tr}(A_2)A_3 \otimes \cdots \otimes A_n$ on product elements and extended by linearity.

We consider $\rho = |\psi_1\rangle\langle\psi_2|$ and $A = |\eta_1\rangle\langle\eta_2|$ without loss of generality and write:

$$\mathrm{tr}\left[(\rho \otimes \sigma)(|\Phi_x\rangle\langle\Phi_x| \otimes U_x^\dagger A U_x)\right] = \qquad (3.47)$$

$$= \langle\psi_2 \otimes \Psi|\Phi_x \otimes U_x^\dagger\eta_1\rangle\langle\Phi_x \otimes U_x^\dagger\eta_2|\psi_1 \otimes \Psi\rangle.$$

Using the fact that $(A \otimes \mathbb{I})\Psi = (\mathbb{I} \otimes A^T)\Psi$ for any linear operator A on $\mathbb{C}^n$, where the transposition is defined w.r.t. the orthonormal basis $\{\varphi_i\}_i$, we have:

$$\langle\psi_2 \otimes \Psi|\Phi_x \otimes U_x^\dagger\eta_1\rangle = \langle\psi_2 \otimes \Psi|(U_x \otimes \mathbb{I})\Psi \otimes U_x^\dagger\eta_1\rangle$$

$$= \langle\psi_2 \otimes (\mathbb{I} \otimes U_x)\Psi|(U_x \otimes \mathbb{I})\Psi \otimes \eta_1\rangle$$

$$= \langle(\mathbb{I} \otimes U_x^T \otimes \mathbb{I})\psi_2 \otimes \Psi|(\mathbb{I} \otimes U_x^T \otimes \mathbb{I})\Psi \otimes \eta_1\rangle$$

$$= \langle\psi_2 \otimes \Psi|\Psi \otimes \eta_1\rangle = \frac{1}{n}\langle\psi_2|\eta_1\rangle.$$

Repeating the calculation for the second factor in the right-hand term of (3.47) we obtain:

$$\mathrm{tr}\left[(\rho \otimes \sigma)(|\Phi_x\rangle\langle\Phi_x| \otimes U_x^\dagger A U_x)\right] = \frac{1}{n^2}\langle\psi_2|\eta_1\rangle\langle\eta_2|\psi_1\rangle = \frac{1}{n^2}\mathrm{tr}(\rho A), \quad (3.48)$$

then (3.46) is proved by summation. Therefore $T^*(\rho \otimes \sigma) = \rho$ for any $\rho \in \mathcal{B}^*(H)$.

(E, L, σ) is a quantum teleportation scheme, so a quantum state can be teleported by means of an entanglement-assisted process. Therefore quantum information can be transmitted through a classical channel once an entangled state has been shared by two clients without the need of sending a quantum system prepared in the target state. Let us conclude this section with a standard example of quantum teleportation that has been experimentally realized in 1997 [5].

Example 3.2 A quantum teleportation scheme of a qubit can be obtained in the following way: The entangled state to be shared is a Bell state:

$$\Phi^+ = \frac{1}{\sqrt{2}}(|11\rangle + |00\rangle),$$

the POVM $\{|\Phi_j\rangle\langle\Phi_j|\}_{j=0,1,2,3}$ describing a measurement on *target+ancilla* is defined by the four Bell states. The parameter-dependent operation is given by the family $\{U_j\}_{j=0,1,2,3}$ of orthogonal unitary operators defined by the identity and the three Pauli matrices:

$$U_0 := \mathbb{I} \quad \text{and} \quad U_j := \sigma_j \quad \text{for} \quad j = 1, 2, 3 \qquad (3.49)$$

3.4.4 Dense Coding

Now let us consider a transmission procedure of classical information by means of a quantum channel and argue if it can be advantageous.

Let $x \in X = \{1, 2, 3, \ldots, n\}$ be the classical datum to be transmitted. Suppose Alice prepares a n-level quantum system in $\rho_x \in \mathfrak{S}(\mathsf{H})$ and sends it to Bob which performs a measurement described by the POVM $\{E_y\}_{y \in Y}$ with $Y = \{1, 2, \ldots, m\}$. The probability that Bob reads the classical datum $y \in Y$ on his device when Alice transmits $x \in X$ is:

$$\mathfrak{P}(y|x) = \text{tr}(\rho_x E_y). \tag{3.50}$$

Let us recall that a classical channel $T : \mathscr{C}(Y) \to \mathscr{C}(X)$ is completely described by its matrix elements $\{T_{xy}\}_{x \in X, y \in Y}$ (3.35) as every linear map, T_{xy} is interpreted as the probability that the output is y when the input is x. Then $T_{xy} := \{\mathfrak{P}(y|x)\}_{x \in X, y \in Y}$ defines a classical channel, denoted by T. More precisely the action of T^* on the probability distribution (i.e. the classical state) $p \in \mathfrak{S}(\mathscr{C}(X))$ yields the probability distribution $T^* p = q \in \mathfrak{S}(\mathscr{C}(Y))$:

$$q(y) = \sum_{x \in X} p(x) \text{tr}(\rho_x E_y). \tag{3.51}$$

The simplest notion of quantum coding of classical information is the following: Given an orthonormal basis $\{\psi_i\}_{i=1,\ldots,n}$ of H we define the parameter-dependent preparation $X \ni x \mapsto \rho_x := |\psi_x\rangle\langle\psi_x|$ (coding) and the measurement process $X \ni y \mapsto E_y := |\psi_y\rangle\langle\psi_y|$ (reading). However we do not take any advantage from such a quantum transmission of classical information in terms of *capacity* or *security* w.r.t. a classical transmission itself.

Exploiting entanglement as a resource we can define a coding procedure where a single qubit can be used to transmit the information of two classical bits. The *Holevo theorem* [12] implies that one cannot convey more than n classical bits of information in n qubit, however an initial entangled state shared between the clients allows to transfer two bits via one qubit.

Ingredients of a dense coding scheme are: a parameter-dependent operation $\{L_x\}_{x \in X}$, where $L_x : \mathfrak{B}(\mathsf{H}) \to \mathfrak{B}(\mathsf{H})$, an entangled state $\sigma \in \mathfrak{S}(\mathsf{H} \otimes \mathsf{H})$ and a POVM measurement $\{E_y\}_{y \in Y} \subset \mathfrak{B}(\mathsf{H} \otimes \mathsf{H})$.

Suppose a composite system is prepared in σ, then a subsystem is sent to Alice and the other to Bob. Assume Alice wants to transmit the datum $x \in X$, she performs quantum operation $L_x : \mathfrak{B}(\mathsf{H}) \to \mathfrak{B}(\mathsf{H})$ on her system. She sends her system to Bob which performs measurements $\{E_y\}_{y \in Y}$ on the bipartite system in his hands. The probability that Bob obtains the outcome $y \in Y$ is given by:

$$\mathfrak{P}(y|x) = \text{tr}[(L_x^* \otimes id)(\sigma)E_y] \equiv T_{xy}. \tag{3.52}$$

Definition 3.8 Let $T = \{T_{xy}\}_{x,y}$ be the classical channel defined in (3.52). If T is ideal (i.e. $X = Y$ and $T_{xy} = \delta_{xy}$) then the triple (E, L, σ) is called **dense coding scheme**.

Let us discuss how a dense coding scheme works in detail. Let H be a n-dimensional Hilbert space and $X = \{0, 1, 2, \ldots, n^2 - 1\}$, let $\sigma = |\Psi\rangle\langle\Psi| \in \mathfrak{S}(H \otimes H)$ be the maximally entangled state defined in (3.43), let $\{U_x\}_{x \in X}$ be a family of unitary operators in H such that $\operatorname{tr}(U_x^\dagger U_y) = n\,\delta_{xy}$ and $\{\Phi_x\}_{x \in X}$ be the orthonormal basis of $H \otimes H$ defined by $\Phi_x := (U_x \otimes \mathbb{I})\Psi$.

Proposition 3.5 *Consider the POVM $\{E_x\}_{x \in X}$ defined by $E_x := |\Phi_x\rangle\langle\Phi_x|$ and the parameter-dependent operation $\{L_x\}_{x \in X}$ defined by $L_x(A) := U_x^\dagger A U_x$. The triple (E, L, σ) is a dense coding scheme.*

Proof Let us prove the following identity:

$$\operatorname{tr}[(L_x^* \otimes \operatorname{id})(\sigma)E_y] = \delta_{xy}, \tag{3.53}$$

which implies the claim. By direct calculation on the left-hand side of (3.53):

$$\operatorname{tr}[(L_x^* \otimes \operatorname{id})(\sigma)E_y] = \langle\Phi_y|(L_x^* \otimes \operatorname{id})\Psi\rangle\langle\Psi|\Phi_y\rangle = |\langle(U_y \otimes \mathbb{I})\Psi|(U_x \otimes \mathbb{I})\Psi\rangle|^2,$$

where we used the fact that $\Phi_y = (U_y \otimes \mathbb{I})\Psi$ and $(L_x^* \otimes \operatorname{id})|\Psi\rangle\langle\Psi| = |(U_x \otimes \mathbb{I})\Psi\rangle\langle(U_x \otimes \mathbb{I})\Psi|$.

Since Ψ is the maximally entangled states defined in (3.43), we have:

$$\langle(U_y \otimes \mathbb{I})\Psi|(U_x \otimes \mathbb{I})\Psi\rangle = \left\langle \frac{1}{\sqrt{n}}\sum_{i=1}^n U_y\varphi_i \otimes \psi_i \,\middle|\, \frac{1}{\sqrt{n}}\sum_{j=1}^n U_y\varphi_j \otimes \psi_j \right\rangle =$$

$$= \frac{1}{n}\sum_{i,j=1}^n \langle U_y\varphi_i|U_x\varphi_j\rangle\langle\psi_i|\psi_j\rangle = \frac{1}{n}\sum_{i=1}^n \langle U_y\varphi_i|U_x\varphi_i\rangle = \delta_{xy}.$$

If $H \simeq \mathbb{C}^2$, $\sigma = \Phi^+$, $\{\Phi_x\}_x$ are the four Bell states and $\{U_x\}_x$ are given by (3.49) then we have a dense coding scheme for a qubit where $X = \{0, 1, 2, 3\} \equiv \{00, 01, 10, 11\}$ so a single qubit transmission carries the information of two classical bits (*superdense coding*). One might object that if Alice prepares the Bell pair in Φ^+ in her lab, she sends one qubit to Bob in order to obtain the initial entanglement sharing and she sends another qubit to transmit the bit-pair then Alice must transmit two qubits anyway. However no information is sent transmitting half of the entangled pair, the classical information (bit-pair) is communicated only by the transmission of the second qubit. Therefore an eavesdropper which intercepts such a qubit cannot gain the classical information that is locked by the initial entanglement sharing.

3.5 Final Remarks and Perspectives

In this paper we have given a glance to some elementary applications of quantum entanglement as a resource in communication schemes. In particular we have introduced and discussed the quantum channels formalism (CP-maps and some of their representations) that is crucial to describe the considered quantum processes within a solid mathematical framework.

Let us point out that entanglement is a well-known resource also for *quantum key distribution* where violation of Bell inequalities can be observed to certify the security of a quantum channel during the sharing of a private key [19]. However entanglement is not the only kind of quantum correlations and *quantum discord* has been introduced in [17] to quantify general quantum correlations contained in a state. A review on quantum discord as a resource for private communications is given in [11].

Beyond the topics of this work quantum channels are important to describe the presence of *quantum noise*, as mentioned in Sect. 3.3.3, *depolarizing channel* and *phase dumping* are celebrated examples. Therefore this formalism is used in the context of quantum error-correction and to formulate security proofs for quantum cryptographic protocols.

Moreover quantum channels give a model to describe non-unitary evolution of open quantum systems in a more general scenario w.r.t. the *master equation approach* in the Lindblad form. In fact once solved the master equation one can use the time dependence of the considered density matrix to define the Kraus operators of a quantum channel, on the other hand the Kraus representation of a quantum channel is not necessarily given in terms of the solution of a master equation. However there are examples of quantum processes that cannot be described by quantum channels [8], thus matter for future research works will be a description of quantum processes generalizing the CP-maps framework.

About recent works on entanglement characterization and its applications as a resource, there have been several attempts to study quantum entanglement considering the geometrization of quantum mechanics, e.g. exploiting the representation of pure states as points of the projective Hilbert space as a real manifold with a Kähler structure [1, 4, 18]. Goals of further investigations could be definition and characterization of the analogous of quantum channels within the pure geometric scenario in order to reach a definitely more general description of quantum processing.

Acknowledgements I would like to thank the organizers of the workshop *Quantum Physics and Geometry* held in Levico Terme (Trento, Italy), 4–6 July 2017: E. Ballico, A. Bernardi, I. Carusotto, S. Mazzucchi, V. Moretti. I am grateful to the referee of the manuscript for useful suggestions and comments.
This work is supported by:

References

1. P. Aniello, J. Clemente-Gallardo, G. Marmo, G.F. Volkert, Classical tensors and quantum entanglement II: mixed states. Int. J. Geom. Meth. Mod. Phys. **08**, 853–883 (2011)
2. A. Aspect, J. Dalibard, G. Roger, Experimental test of Bell's inequalities using time-varying analyzers. Phys. Rev. Lett. **49**, 1804–1807 (1982)
3. J.S. Bell, On Einstein Podolski Rosen paradox. Physics **1**, 195–200 (1964)
4. I. Bengtsson, K. Zyczkowski, *Geometry of Quantum States. An introduction to Quantum Entanglement* (Cambridge University Press, Cambridge, 2006)
5. D. Bouwmeester, J.W. Pan, K. Mattle, M. Eibl, H. Weinfurter, A. Zeilinger, Experimental quantum teleportation. Nature **390**, 575–579 (1997)
6. P. Busch, No information without disturbance: quantum limitations of measurement, in *Quantum Reality, Relativistic Causality, and Closing the Epistemic Circle* (Springer, Dordrecht, 2009)
7. V. Buzek, M. Hillery, Quantum copying: beyond the no-cloning theorem. Phys. Rev. A **54**, 1844 (1996)
8. I. Chuang, M. Nielsen, *Quantum Computation and Quantum Information* (Cambridge University Press, Cambridge, 2000)
9. E.B. Davies, *Quantum Theory of Open Systems* (Academic, London, 1976)
10. A. Einstein, B. Podolski, N. Rosen, Can quantum-mechanical description of physical reality be considered complete? Phys. Rev. **47**, 777 (1935)
11. M. Gu, S. Pirandola, Discord, quantum knowledge and private communications, in *Lectures on General Quantum Correlations and Their Applications* (Springer International Publishing, Cham, 2017), pp. 231–239
12. A.S. Holevo, Bounds for the quantity of information transmitted by a quantum communication channel. Probl. Inf. Transm. **9**, 177–183 (1973)
13. M. Horodecki, P. Horodecki, R. Horodecki, Separability of mixed states: necessary and sufficient conditions. Phys. Lett. A **223**, 1–8 (1996)
14. M. Keyl, Fundamentals of quantum information theory. Phys. Rep. **369**(5), 431–548 (2002)
15. G. Lindblad, A general no-cloning theorem. Lett. Math. Phys. **47**(2), 189–196 (1999)
16. V. Moretti, *Spectral Theory and Quantum Mechanics* (Springer, Milan, 2013)
17. H. Ollivier, W.H. Zurek, Quantum discord: a measure of quantumness of correlations. Phys. Rev. Lett. **88**, 017901 (2001)
18. D. Pastorello, A geometric Hamiltonian description of composite quantum systems and quantum entanglement. Int. J. Geom. Meth. Mod. Phys. **12**, 1550069 (2015)
19. D. Pastorello, A quantum key distribution scheme based on tripartite entanglement and violation of CHSH inequality. Int. J. Quantum Inf. **15**, 1750040 (2017)
20. A. Peres, Separability criterion for density matrices. Phys. Rev. Lett. **77**, 1413–1415 (1996)
21. M.B. Plenio, S. Virmani, An introduction to entanglement measures. J. Quantum Inf. Comput. **7**, 1–51 (2007)

22. W.F. Stinespring, Positive functions on C^*-algebras. Proc. Am. Math. Soc. **6**, 211–216 (1955)
23. F. Strocchi, *An Introduction to the Mathematical Structure of Quantum Mechanics*, 2nd edn. (World Scientific Publishing, Singapore, 2008)
24. R.F. Werner, All teleportation and dense coding schemes. J. Phys. A **34**(35), 7081 (2001)
25. W.K. Wootters, W.H. Zurek, A single quantum cannot be cloned. Nature **299**, 802–803 (1982)

Chapter 4
Frontiers of Open Quantum System Dynamics

Bassano Vacchini

Abstract We briefly examine recent developments in the field of open quantum system theory, devoted to the introduction of a satisfactory notion of memory for a quantum dynamics. In particular, we will consider a possible formalization of the notion of non-Markovian dynamics, as well as the construction of quantum evolution equations featuring a memory kernel. Connections will be draw to the corresponding notions in the framework of classical stochastic processes, thus pointing to the key differences between a quantum and classical formalization of the notion of memory effects.

4.1 Introduction

The theory of open quantum systems denotes the application of quantum mechanics to situations in which the dynamics of the quantum system of interest is influenced by other degrees of freedom, that we are neither interested nor capable to take into account in detail. While in principle any system has to be considered open, since a perfect shielding from the environment is never feasible, in simple situations isolation can be considered as a good approximation. In many realistic settings however, the effect of an external quantum environment on the system dynamics cannot be neglected. This is the case e.g. in many instances of quantum optics, condensed matter physics and quantum chemistry. For the case of an open system dynamics, many interesting physical effects and mathematical structures do appear [1]. From the physical viewpoint, with respect to a closed dynamics we are faced with new phenomena like dissipation and decoherence, which only have partial analog in the classical setting. Such phenomena play a crucial role in many

B. Vacchini (✉)
Dipartimento di Fisica "Aldo Pontremoli", Università degli Studi di Milano, Milan, Italy

INFN, Sezione di Milano, Milan, Italy
e-mail: bassano.vacchini@mi.infn.it

© Springer Nature Switzerland AG 2019
E. Ballico et al. (eds.), *Quantum Physics and Geometry*,
Lecture Notes of the Unione Matematica Italiana 25,
https://doi.org/10.1007/978-3-030-06122-7_4

relevant recent fields of research, such as quantum computation and quantum thermodynamics [2–4].

In this research field many problems are still open, which have important connections to mathematics. In this contribution we will try to highlight, in a concise way, some recent developments in the field of open quantum systems, connected to some of the presently most active research lines, relevant for the mathematical formulation of the theory. After a brief description of the framework of open quantum systems in Sect. 4.2, we will address in Sect. 4.3 the delicate question of the definition of non-Markovian quantum processes. This implies introducing a notion of non-Markovian dynamics, which is quite different form the classical one, though the two can be naturally connected. In Sect. 4.4 we will point to the derivation of equations of motions allowing to introduce memory effects, focusing in particular on master equations with a memory kernel, for which again a natural connection to a class of non-Markovian classical processes can be considered.

4.2   Open Quantum System Dynamics

Let us first introduce some basic elements of open quantum system theory [1], which actually consists in considering the dynamics of a quantum system described on the Hilbert space $\mathcal{H}_S$ without assuming it to be isolated, so that it interacts with an external environment described on a Hilbert space $\mathcal{H}_E$ by means of unitary operators $U(t)$ acting on $\mathcal{H}_S \otimes \mathcal{H}_E$. The fact that the system is not closed brings with itself two important new aspects. On the one hand, the reduced dynamics of the isolated system only is not described by a Liouville von-Neumann equation, and purity of the state is not preserved during the evolution. On the other hand, even a factorized system-environment state develops correlations, so that the latter play a major role in the time evolution. The tensor product structure of the underlying Hilbert space, on its turn, brings in two important aspects. On the one hand, states can exhibit correlations which are of non-classical nature, such as entanglement. On the other hand, the evolution of the reduced system as a function of time is described by a collection of transformations which have the property of being completely positive, a property strictly connected to the non-commutativity of the space of observables. Assuming that the state at the initial time is factorized

$$\rho_{SE}(0) = \rho_S(0) \otimes \rho_E, \tag{4.1}$$

we have that the reduced state of the system at a later time is given by

$$\rho_S(t) = \mathrm{Tr}_E\{U(t)\rho_S(0) \otimes \rho_E U(t)^\dagger\}, \tag{4.2}$$

where Tr_E denotes the partial trace with respect to the environmental degrees of freedom. This state contains all the information relevant for the description of the dynamics of the system observables. In particular this transformation defines a linear map

$$\Phi(t, 0)[\rho_S(0)] \equiv \mathrm{Tr}_E\{U(t)\rho_S(0) \otimes \rho_E U(t)^\dagger\}, \tag{4.3}$$

which considering an orthogonal resolution for the state of the environment $\rho_E = \sum_\xi \lambda_\xi P_{\varphi_\xi}$, and introducing an orthogonal basis $\{\varphi_\eta\}$ in $\mathcal{H}_E$, admits the following representation

$$\Phi(t, 0)[\rho_S(0)] = \sum_{\xi,\eta} \lambda_\xi \langle \varphi_\eta | U(t)\varphi_\xi \rangle \rho_S(0)(\langle \varphi_\eta | U(t)\varphi_\xi \rangle)^\dagger$$

$$= \sum_{\xi,\eta} K_{\xi\eta}(t)\rho_S(0)K_{\xi\eta}^\dagger(t), \tag{4.4}$$

where we have introduced so-called Kraus operators $K_{\xi\eta} = \sqrt{\lambda_\xi} \langle \varphi_\eta | U(t)\varphi_\xi \rangle$ acting on $\mathcal{H}_S$. This representation warrants complete positivity of the map. A map Φ defined on the space of trace class operators $\mathcal{T}(\mathcal{H}_S)$ is said to be completely positive if its extension $\Phi \otimes \mathbb{1}_n$ to $\mathcal{T}(\mathcal{H}_S \otimes \mathbb{C}^n)$ defined on operators in tensor product form as

$$\Phi \otimes \mathbb{1}_n[A \otimes B] = \Phi[A] \otimes B$$

is a positive map for any $n \in \mathbb{N}$. Otherwise stated, the trivial extension of a completely positive map acting on some system, to a larger set of degrees of freedom the system is not interacting with, remains positive. It can be shown that any such map admits the representation Eq. (4.4), and viceversa [5]. For an initial state in factorized form as in Eq. (4.1) it is thus possible to define a reduced dynamics, described by the time dependent collection of completely positive trace preserving maps $\Phi(t, 0)$ given by Eq. (4.3), as shown in Fig. 4.1. Two natural questions appear at this stage. On the one hand, now that reversibility of the dynamics warranted by the unitary evolution has got lost, it is interesting to ascertain whether such maps do describe memory effects. On the other hand, one would like to know the possible expression of maps $\Phi(t, 0)$ describing a well defined dynamics, as well as the

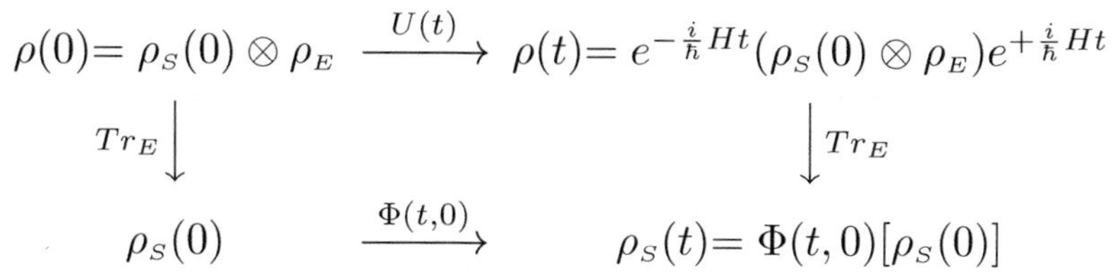

$$\rho(0) = \rho_S(0) \otimes \rho_E \xrightarrow{\;\;U(t)\;\;} \rho(t) = e^{-\frac{i}{\hbar}Ht}(\rho_S(0) \otimes \rho_E)e^{+\frac{i}{\hbar}Ht}$$

$$Tr_E \downarrow \qquad\qquad\qquad\qquad\qquad\qquad \downarrow Tr_E$$

$$\rho_S(0) \xrightarrow{\;\;\Phi(t,0)\;\;} \rho_S(t) = \Phi(t, 0)[\rho_S(0)]$$

Fig. 4.1 Commutative diagram showing the existence of a reduced dynamics for an initial system-environment state in factorized from. The reduced state of the system at time t can be equivalently obtained by taking the marginal with respect to the environmental degrees of freedom of the unitarily evolved total state, or by applying the completely positive trace preserving map $\Phi(t, 0)$ to the initial state of the system

general structure of evolution equations for the statistical operator admitting such collection of maps as solution. These two aspects have been the object of extensive research, and some recent developments in this respect will be discussed Sects. 4.3 and 4.4.

4.3   Characterization of Dynamics with Memory

The existence of the reduced dynamics for an open quantum system implies that its time evolution can be described by evolution equations which, on top of a coherent quantum dynamics as can be obtained by a Liouville-von Neumann equation, do exhibit stochasticity. The stochastic contribution to the dynamics is due to the interaction with the unobserved quantum degrees of freedom of the environment. A quite natural question in this setting, in analogy with what happens for a classical stochastic dynamics, is therefore whether the obtained quantum dynamics can exhibit effects which can be reasonably termed memory effects. In the description of classical systems random features in the dynamics are described by the mathematically well established notion of stochastic process. The notion of lack of memory for a stochastic process is enforced by asking a suitable constraint on the conditional probabilities determining the process. Roughly speaking, a process is defined to be Markovian, that is without memory, if the only relevant conditioning of the probability densities for the outcomes of the considered stochastic process is with respect to the last ascertained value of the time dependent random variable considered, and not with respect to values at previous times. In such a way a notion of memory is naturally introduced (see e.g. [6] for a proper formalization of this notion). For a Markov process the notion of lack of memory is therefore naturally linked to the neglecting of knowledge of values taken by the random variable in the past.

   In the quantum framework, random variables have to be described by self-adjoint operators acting in the Hilbert space $\mathcal{H}_S$ of the considered system. However, in order to obtain the probability distribution for the values taken by such random variables at a given time one has to perform a measurement. At variance with the classical case, knowledge of the value of the random variable will thus affect the subsequent evolution of the system in a non negligible way, depending on the way the measurement is performed. The external intervention necessary for the measurement thus influences the value of multitime probability densities, which do not admit an obvious definition as in the classical case. The definition of a quantum Markovian process along the lines of the classical viewpoint, pursued in the first systematic studies on the characterisation of open quantum system dynamics [7–9], thus encounters major difficulties.

   More recently, different approaches have been considered, which tackle the issue considering features of the dynamics determined by quantities depending on a single point in time, rather than on multitime probability densities (see [10–13] for reviews). In such a way one can overcome the difficulties related to the measurement problem and allow for a direct experimental verification of the property of Markovianity. The connection between these approaches and the notion of classical Markovian process has been discussed in [14]. In this contribution we will concentrate on a approach which connects non-Markovianity to a reversible exchange of information between system and environment, started with the seminal paper [15]. Let us consider a reduced dynamics on $\mathcal{H}_S$ defined by a collection of completely positive trace preserving maps $\Phi(t, 0)$ sending the initial system state to the state at time t

$$\Phi \equiv \{\Phi(t, 0)\}_{t \in \mathbb{R}_+},$$

and suppose that the experimenter can prepare two distinct initial states of the system, say $\rho_S^1(0)$ and $\rho_S^2(0)$, with the same probability $p = 1/2$. If an observer has to guess which state has actually been prepared by performing a single measurement, as shown in [16] the maximal probability of success, obtained by performing an optimal measurement, is given by the expression

$$P(0) = \frac{1}{2}(1 + D(\rho_S^1(0), \rho_S^2(0))), \tag{4.5}$$

where

$$D(\rho, \sigma) = \frac{1}{2}\|\rho - \sigma\|_1$$

denotes the trace distance between two statistical operators $\rho, \sigma \in \mathcal{T}(\mathcal{H}_S)$, namely the normalized distance built by means of the trace norm $\|\cdot\|_1$. If the observer tries to distinguish the states at a later time t, after interaction of the system with the environment, the success probability is now given by

$$P(t) = \frac{1}{2}(1 + D(\rho_S^1(t), \rho_S^2(t))), \tag{4.6}$$

where $\rho_S^{1,2}(t) = \Phi(t, 0)[\rho_S^{1,2}(0)]$ and due to a crucial property of the trace distance we have

$$P(t) \leqslant P(0).$$

Indeed the trace distance is a contraction under the action of an arbitrary positive, and therefore in particular completely positive, trace preserving transformation Λ [17]

$$D(\Lambda[\rho], \Lambda[\sigma]) \leqslant D(\rho, \sigma).$$

The effect of the interaction with the environment is thus a reduction of the capability to distinguish quantum states of the system by an observer performing measurements on the system only. For the case in which the dynamics is characterised by a semigroup composition law so that

$$\Phi(t,0) = e^{t\mathcal{L}}, \tag{4.7}$$

with $\mathcal{L}$ a suitable generator, corresponding to dynamics typically called quantum Markov processes, one further has

$$P(t) \leqslant P(s) \qquad \forall t \geqslant s,$$

so that there is a monotonic decrease in time of the distinguishability between system states. This feature is taken as the defining property of a quantum Markovian dynamics. Accordingly, a quantum dynamics described by a collection of completely positive trace preserving maps $\Phi(t,0)$ is said to be non-Markovian if there are revivals in time in the success probability $P(t)$ or equivalently in the trace distance $D(\rho_S^1(t), \rho_S^2(t))$, which contains the relevant part of the information, so that

$$\dot{D}(\rho_S^1(t), \rho_S^2(t)) \geqslant 0, \tag{4.8}$$

for at least a point in time and a couple of initial states, where we have denoted by $\dot{D}$ the time derivative of the trace distance between the evolved initial system states.

These revivals do generally depend on the choice of initial states, so that a suitable quantifier of non-Markovianity of the dynamics has been introduced according to the expression

$$\mathcal{N}(\Phi) = \max_{\rho_S^{1,2}(0)} \int_{\dot{D}>0} dt \, \dot{D}(\rho_S^1(t), \rho_S^2(t)). \tag{4.9}$$

It immediately appears that this definition of non-Markovian quantum dynamics only requires to observe the state of the system at different times and starting from different system initial conditions, rather than on multitime quantities, so that an experimental assessment of non-Markovianity can be obtained by means of a tomographic procedure [18].

To substantiate the interpretation of this notion of non-Markovianity as information back flow from the environment to the system, let us introduce the following quantities [2, 12]

$$\mathcal{I}_{\mathrm{int}}(t) = D(\rho_S^1(t), \rho_S^2(t)) \tag{4.10}$$

and

$$\mathcal{I}_{\mathrm{ext}}(t) = D(\rho_{SE}^1(t), \rho_{SE}^2(t)) - D(\rho_S^1(t), \rho_S^2(t)), \tag{4.11}$$

where $\mathcal{I}_{\text{int}}(t)$ is used to quantify the internal information, that is the information accessible by performing measurements on the system only, while $\mathcal{I}_{\text{ext}}(t)$ denotes the external information, which can only be obtained by performing measurements in the Hilbert space of both system and environment $\mathcal{H}_S \otimes \mathcal{H}_E$, minus the internal one. If the overall dynamics is unitary the sum of the two quantities $\mathcal{I}_{\text{tot}}(t) = \mathcal{I}_{\text{int}}(t) + \mathcal{I}_{\text{ext}}(t)$ is conserved, so that in particular

$$\frac{\mathrm{d}}{\mathrm{d}t} D(\rho_S^1(t), \rho_S^2(t)) = \frac{\mathrm{d}}{\mathrm{d}t}\mathcal{I}_{\text{int}}(t)$$

$$= -\frac{\mathrm{d}}{\mathrm{d}t}\mathcal{I}_{\text{ext}}(t).$$

This equality shows that an increase in time of the trace distance corresponds to a decrease in the external information, which being overall conserved can only flow from the environment into the system. To understand in which sense information can be stored outside the system, namely it cannot be retrieved by performing measurements on the system only, it is enlightening to consider the following bound, first introduced in Ref. [19] in connection to detection of initial correlations

$$D(\rho_S^1(t), \rho_S^2(t)) - D(\rho_S^1(s), \rho_S^2(s)) \leqslant D(\rho_{SE}^1(s), \rho_S^1(s) \otimes \rho_E^1(s)) \quad (4.12)$$

$$+ D(\rho_{SE}^2(s), \rho_S^2(s) \otimes \rho_E^2(s))$$

$$+ D(\rho_E^1(s), \rho_E^2(s)),$$

where it is assumed that $t \geqslant s$ and, at variance with [19], $\rho_{SE}^{1,2}(0) = \rho_S^{1,2}(0) \otimes \rho_E(0)$, where $\rho_S^{1,2}(0)$ and $\rho_E(0)$ are the marginal states obtained by taking the partial trace with respect to the degrees of freedom of environment and system respectively, so as to ensure the existence of a reduced dynamics. As discussed above the trace distance can be naturally understood as a quantifier of distinguishability among quantum states, so that in particular, if the two statistical operators are a state on a bipartite space and the product of its marginals as in the r.h.s. of Eq. (4.12), it provides a quantifier of correlations in the overall state. The l.h.s. corresponds to the change over time in trace distance, which can only be positive if at least one of the quantities at the r.h.s. is different from zero, that is after interacting for a time s either system and environment have become correlated or the environmental state has changed in different ways depending on the initial system state.

This definition of non-Markovianity of a quantum dynamics based on the notion of information back flow between system and environment is strictly connected to an alternative notion relying on a mathematical property of the collection of completely positive trace preserving maps describing the reduced dynamics. Indeed such a collection is called P-divisible if the following identity holds [20]

$$\Phi(t, 0) = \Phi(t, s)\Phi(s, 0) \qquad \forall t \geqslant s \geqslant 0, \quad (4.13)$$

with $\Phi(t, s)$ positive maps for any $t \geqslant s \geqslant 0$, while it is called CP-divisible if the maps $\Phi(t, s)$ are in particular completely positive for any $t \geqslant s \geqslant 0$, as in the case e.g. of the quantum dynamical semigroup considered in Eq. (4.7). It immediately appears that both CP-divisible and P-divisible are Markovian according to the trace distance criterion defined above, while a monotonic decrease of the trace distance in general does not warrant neither kind of divisibility. The composition law Eq. (4.13) tells us that, in order to predict the time evolution of the system forward in time, we only need to know the state at a given time, thus naturally inducing a formalization of lack of memory, and indeed its violation was proposed as a definition of non-Markovian dynamics in [21].

4.3.1 Generalized Non-Markovianity Measure

Recently different refinements of the definition and quantification of non-Markovianity as given by formulae Eqs. (4.8) and (4.9) have been considered [22–24], importantly always supporting the seminal interpretation of non-Markovianity as information back flow from environment to system. An important and natural generalization, first suggested in [25], consists in considering the discrimination problem between quantum states, used to connect trace distance and distinguishability, in the more general setting in which the two known states $\rho_S^1(0)$ and $\rho_S^2(0)$ can be prepared with different weights, say p_1 and p_2. In this case the optimal strategy can be shown to lead to the following success probability

$$P(0) = \frac{1}{2}(1 + \Delta(\rho_S^1(0), \rho_S^2(0); p_1, p_2)), \qquad (4.14)$$

where the expression $\Delta(\rho_S^1(0), \rho_S^2(0); p_1, p_2) = \|p_1\rho_S^1(0) - p_2\rho_S^2(0)\|$ is also known as norm of the Helstrom matrix. Non-Markovianity is then identified with a revival in time of the norm of the Helstrom matrix

$$\dot{\Delta}(\rho_S^1(t), \rho_S^2(t); p_1, p_2) \geqslant 0, \qquad (4.15)$$

for at least a point in time, a couple of initial states and a choice of weights, which provide apriori information on the prepared state. Note that the class of processes which are non-Markovian is thus enlarged, including situations which where previously not encompassed [23, 26]. Accordingly, a generalized measure of non-Markovianity can be considered, given by the expression

$$\mathcal{N}(\Phi) = \max_{p_{1,2}, \rho_S^{1,2}(0)} \int_{\dot{\Delta} > 0} dt \, \dot{\Delta}(\rho_S^1(t), \rho_S^2(t); p_1, p_2). \qquad (4.16)$$

An important result of this generalization of the initial definition is the fact that it allows for a clearcut connection with the notion of divisibility considered in

Eq. (4.13), which in its mathematical formulation does not immediately show a link to information flow, since the latter can only be formulated by introducing a quantifier of distinguishability among states. Indeed, thanks to a result by Kossakowski connecting the positivity property of a trace preserving map with its contractivity when acting on an arbitrary hermitian observable [27], monotonicity in time of the behavior of the norm of the Helstrom matrix $\Delta(\rho_S^1(t), \rho_S^2(t); p_1, p_2)$ can be shown to be equivalent to P-divisibility of the collections of time evolution maps in the sense of Eq. (4.13), provided the time evolution map is invertible as a linear map on the space of operators. Most importantly, this extension is still compatible with the notion of information back flow as characterizing a non-Markovian dynamics. This fact can be shown considering suitable generalizations of the notion of internal and external information as considered in Eq. (4.10) and Eq. (4.11), as well as a generalisation of the bound Eq. (4.12), thus pointing to the general validity of the starting definition of quantum non-Markovianity [24, 28].

4.4   Non-Markovian Evolution Equations

As discussed in Sect. 4.2, considering an initially factorized state of system and environment is sufficient to warrant the existence of a reduced dynamics, which will depend both on the state of the environment and the unitary interaction, according to expression Eq. (4.2). However, in the general case the evaluation of the exact dynamics is utterly unfeasible, so that it is of utmost importance to have access to approximate methods. On the one hand, one can consider perturbation expansions; on the other hand, one can look for phenomenological expressions. In both cases one major difficulty is warranting that the obtained time evolutions indeed provide a well-defined dynamics, corresponding to a completely positive trace preserving transformation. In particular, the requirement of complete positivity, which warrants connection to an underlying microscopic dynamics, is difficult to be enforced and is typically lost at intermediate steps in a perturbative approach. A fundamental result has been obtained for the situation in which the time evolution, instead of obeying a group evolution law as in the case of a reversible unitary dynamics, can be described by a semigroup, thus introducing a preferred direction in time. In this case the collection of maps is called quantum dynamical semigroup and is determined by a generator $\mathcal{L}$ according to Eq. (4.7). A fundamental theorem of open quantum system theory [14, 15] states that this generator has to be in the so-called Gorini-Kossakowksi-Sudarshan-Lindblad form

$$\mathcal{L}[\rho_S] = -\frac{i}{\hbar}[H, \rho_S] + \sum_k \gamma_k \left[L_k \rho_S L_k^\dagger - \frac{1}{2}\{L_k^\dagger L_k, \rho_S\} \right], \qquad (4.17)$$

with H a self-adjoint operator corresponding to an effective Hamiltonian, γ_k positive rates and L_k system operators also called Lindblad operators. It provides a generalization of the Liouville von-Neumann equation to include both decoherence and dissipative effects. Solutions of the time evolution equation

$$\frac{\mathrm{d}}{\mathrm{d}t}\rho_S(t) = \mathcal{L}[\rho_S(t)] \tag{4.18}$$

together with a suitable initial condition $\rho_S(0)$ do define a collection of completely positive trace preserving maps obeying a semigroup composition law. As discussed in Sect. 4.3, the obtained dynamics is Markovian and provides the quantum analog of a classical semigroup evolution. To describe memory effects more general dynamics have to be considered. To this aim one can either consider time dependent generalizations of the generator considered in Eq. (4.17), or move to evolution equations explicitly featuring a memory kernel. In both cases one has to ensure that the solutions of such equations do provide a collection of time dependent completely positive trace preserving maps, thus describing a well defined dynamics. In the case of so-called time local evolution equations, one has to replace rates and operators appearing in Eq. (4.17) by time dependent quantities, looking for conditions warranting complete positivity. Considering master equations in integrodifferential form

$$\frac{\mathrm{d}}{\mathrm{d}t}\rho_S(t) = \int_0^t \mathrm{d}\tau \mathcal{K}(t-\tau)[\rho_S(\tau)] \tag{4.19}$$

the corresponding task is to envisage conditions on the operator kernel $\mathcal{K}(t)$ warranting preservation of positivity and trace of the solutions of the integrodifferential equation. In both cases the most general solution to the problem is not known, even not heuristically, while partial results have been recently obtained [29–36]. In particular we will consider how to obtain well-defined quantum memory kernels $\mathcal{K}(t)$. While quantum dynamical semigroups can be seen as the quantum counterpart of classical Markov semigroups, the class of considered memory kernels can be taken as the quantum analogue of a class of non-Markovian processes known as semi-Markov process [37].

We consider as starting point an expression for the exact solution of Eq. (4.17), which can be written as

$$\rho_S(t) = \Phi(t, 0)[\rho_S(0)] \tag{4.20}$$

$$= \mathcal{R}(t)[\rho_S(0)]$$

$$+ \sum_{k=1}^{\infty} \int_0^t \mathrm{d}t_k \ldots \int_0^{t_2} \mathrm{d}t_1\, \mathcal{R}(t-t_k)\mathcal{J}\ldots\mathcal{R}(t_2-t_1)\mathcal{J}\mathcal{R}(t_1)[\rho_S(0)],$$

where we have introduced the contraction semigroup

$$\mathcal{R}(t)[\rho] = \mathrm{e}^{-\frac{i}{\hbar}Ht - \frac{1}{2}\sum_k \gamma_k L_k^\dagger L_k t} \rho \mathrm{e}^{+\frac{i}{\hbar}Ht - \frac{1}{2}\sum_k \gamma_k L_k^\dagger L_k t}$$

and the completely positive map

$$\mathcal{J}[\rho] = \sum_k \gamma_k L_k \rho L_k^{\dagger}.$$

As a result the solution is expressed as a sum of contributions characterised by a given number of insertions of the completely positive map $\mathcal{J}$ with an intermediate trace decreasing evolution in between. Complete positivity of the overall evolution is warranted by the fact that we are considering sum and composition of completely positive maps, which form a convex cone. The solution is expressed in a space of "trajectories" determined by the number of jumps or insertions of the map $\mathcal{J}$ and the points in time at which these jumps happen [38, 39]. Both maps $\mathcal{R}(t)$ and $\mathcal{J}$ are determined by the rates γ_k and the Lindblad operators L_k. To consider a more general situation one can define a collection of linear maps in analogy with Eq. (4.20), introducing the replacement of jump operator and contraction semigroup by means of an arbitrary completely positive trace preserving transformation $\mathcal{E}$ and a collection of time dependent completely positive trace preserving maps $\mathcal{F}(t)$, according to the scheme

$$\rho_S(t) = g(t)\mathcal{F}(t)[\rho_S(0)] \tag{4.21}$$

$$+ \sum_{k=1}^{\infty} \int_0^t dt_k \dots \int_0^{t_2} dt_1 \, f(t - t_k)\mathcal{F}(t - t_k)\mathcal{E} \dots$$

$$\times \dots f(t_2 - t_1)\mathcal{F}(t_2 - t_1)\mathcal{E} g(t_1)\mathcal{F}(t_1)[\rho_S(0)].$$

In the representation Eq. (4.21) we have further inserted the functions $f(t)$ and $g(t)$. The function $f(t)$ has to be positive and normalized to one over the interval $[0, \infty)$, so as to be interpreted as a waiting time distribution. Accordingly, the function $g(t)$ is determined by $\dot{g}(t) = -f(t)$, together with $g(0) = 1$, so that it can be interpreted as the associated survival probability. It can be easily seen that these properties are sufficient to identify the linear assignment $\rho_S(0) \rightarrow \rho_S(t)$ obtained through Eq. (4.21) as a collection of completely positive trace preserving maps. These evolutions correspond to a situation in which in between the evolution given by the maps $\mathcal{F}(t)$, the system undergoes a transformation described by the completely positive trace preserving map $\mathcal{J}$. It is however not obvious the existence of a closed evolution equation for the statistical operator of the system $\rho_S(t)$, so as to connect the transformations to a continuous dynamics. To this aim one considers the expression of the Laplace transform of Eq. (4.21), which thanks to the presence of convolutions takes the simple form

$$\widehat{\rho}_S(u) = (\mathbb{1} - \widehat{f\mathcal{F}}(u)\mathcal{E})^{-1}\widehat{g\mathcal{F}}(u)\rho_S(0),$$

where the hat denotes the Laplace transform, so that by a suitable rearrangement one has

$$u\hat{\rho}_S(u) - \rho_S(0) = \left[\frac{1}{\widehat{g\mathcal{F}}(u)}\widehat{f\mathcal{F}}(u)\mathcal{E} - \left(\frac{1}{\widehat{g\mathcal{F}}(u)} - u\right)\right]\hat{\rho}_S(u), \quad (4.22)$$

allowing to identify the memory kernel $\mathcal{K}(t)$ in Eq. (4.19) with the inverse Laplace transform of the operator

$$\widehat{\mathcal{K}}(u) = \frac{1}{\widehat{g\mathcal{F}}(u)}\widehat{f\mathcal{F}}(u)\mathcal{E} - \left(\frac{1}{\widehat{g\mathcal{F}}(u)} - u\right), \quad (4.23)$$

thus showing in particular that indeed the transformation Eq. (4.21) describes a closed dynamics. Actually it can be shown that the kernel Eq. (4.23) despite its complex expression does have a simple and natural interpretation and allows for a connection with a class of non-Markovian processes known as semi-Markov [40, 41]. These generally non-Markovian classical processes describe a dynamics in a discrete state space, in which jumps from site m to site n take place with probability given by the elements of a stochastic matrix π_{nm}, at times distributed according to the waiting time distribution $f_n(t)$. For these processes one can introduce a generalized master equation obeyed by the one-point probability density $P_n(t)$ given by [37, 42]

$$\frac{d}{dt}P_n(t) = \int_0^t d\tau \sum_m [W_{nm}(\tau)P_m(t-\tau) - W_{mn}(\tau)P_n(t-\tau)],$$

whose expression in Laplace transform reads

$$u\hat{P}_n(u) - P_n(0) = \sum_m \left[\pi_{nm}\frac{\hat{f}_m(u)}{\hat{g}_m(u)} - \delta_{nm}\left(\frac{1}{\hat{g}_m(u)} - u\right)\right]\hat{P}_m(u). \quad (4.24)$$

A natural correspondence can be drawn between Eqs. (4.24) and (4.22). The stochastic matrix π_{nm} is replaced by the completely positive trace preserving map $\mathcal{E}$, while the collection of waiting time distributions $f_n(t)$ goes over to $f(t)\mathcal{F}(t)$, product of waiting time distribution and completely positive trace preserving maps. Classical functions are therefore now replaced by operators. The classical dynamics corresponding to jumps between sites with probabilities determined by a given stochastic transition matrix and at times dictated by given waiting time distributions, is replaced by a piecewise quantum dynamics in the space of statistical operators. In this quantum dynamics transformations described by a completely positive trace preserving map $\mathcal{E}$, at times described by a fixed waiting time distribution, are interspersed with a continuous time evolution described by the collection of completely positive trace preserving maps $\mathcal{F}(t)$. It immediately appears that in the correspondence from Eqs. (4.24) to (4.22) an important and typically quantum feature appears, namely the relevance of operator ordering. Indeed Eq. (4.22) can have different quantum counterparts, and another operator ordering leads to an alternative expression for the kernel

$$\widehat{\mathcal{K}}(u) = \mathcal{E}\widehat{f\mathcal{F}}(u)\frac{1}{\widehat{g\mathcal{F}}(u)} - \left(\frac{1}{\widehat{g\mathcal{F}}(u)} - u\right), \qquad (4.25)$$

which substituted in Eq. (4.19) still leads to a well-defined dynamics. Indeed it turns out that the two combinations describe different microscopic modelling of a quantum piecewise dynamics. The microscopic dynamics formalised by Eq. (4.23) corresponds to the physics of the micromaser [43–45], while the kernel Eq. (4.25) naturally appears in so-called collision models [46, 47].

4.5   Conclusions and Outlook

We have briefly exposed recent work within the framework of open quantum system theory, aiming at the definition and quantification of the so-called non-Markovianity, to be understood as the capability of a quantum dynamics to feature memory effects. In particular, we have pointed to a notion of non-Markovian dynamics connected to an information exchange between the considered system and the surrounding environment, whose generalization can be naturally connected to a notion of divisibility of quantum maps. We have further considered a possible extension of a known class of master equations describing a completely positive trace preserving dynamics to include memory effects by means of the introduction of a memory kernel.

Great efforts are presently being put in the endeavour to understand the relevance of the proposed notions of non-Markovian quantum dynamics for the description of relevant physical systems (see in this respect the recent reviews [10–13]). A critical and important open issue is, in particular, whether it captures distinctive features of the dynamics, or if a non-Markovian evolution brings with itself advantages in performing relevant tasks, e.g. in quantum information or quantum thermodynamics.

Acknowledgements The author acknowledges support from the EU Collaborative Project QuProCS (Grant Agreement 641277) and from MIUR through the FFABR project.

References

1. H.-P. Breuer, F. Petruccione, *The Theory of Open Quantum Systems* (Oxford University Press, Oxford, 2002)
2. M. Nielsen, I. Chuang, *Quantum Computation and Quantum Information* (Cambridge University Press, Cambridge, 2000)

3. J. Millen, A. Xuereb, Perspective on quantum thermodynamics. New J. Phys. **18**, 011002 (2016)
4. R. Alicki, R. Kosloff, *Introduction to Quantum Thermodynamics: History and Prospects* (2018). arXiv:1801.08314
5. K. Kraus, *States, Effects, and Operations*. Lecture Notes in Physics, vol. 190 (Springer, Berlin, 1983)
6. D.R. Cox, H.D. Miller, *The Theory of Stochastic Processes* (Wiley, New York, 1965)
7. J.T. Lewis, Quantum stochastic processes 1. Phys. Rep. **77**, 339 (1981)
8. L. Accardi, A. Frigerio, J.T. Lewis, Quantum stochastic processes. Publ. RIMS Kyoto **18**, 97 (1982)
9. G. Lindblad, Non-markovian quantum stochastic processes and their entropy. Commun. Math. Phys. **65**, 281 (1979)
10. H.-P. Breuer, Foundations and measures of quantum non-Markovianity. J. Phys. B **45**, 154001 (2012)
11. A. Rivas, S.F. Huelga, M.B. Plenio, Quantum non-Markovianity: characterization, quantification and detection. Rep. Prog. Phys. **77**, 094001 (2014)
12. H.-P. Breuer, E.-M. Laine, J. Piilo, B. Vacchini, Colloquium : non-Markovian dynamics in open quantum systems. Rev. Mod. Phys. **88**, 021002 (2016)
13. I. de Vega, D. Alonso, Dynamics of non-Markovian open quantum systems. Rev. Mod. Phys. **89**, 015001 (2017)
14. B. Vacchini, A. Smirne, E.-M. Laine, J. Piilo, H.-P. Breuer, Markovianity and non-Markovianity in quantum and classical systems. New J. Phys. **13**, 093004 (2011)
15. H.-P. Breuer, E.-M. Laine, J. Piilo, Measure for the degree of non-Markovian behavior of quantum processes in open systems. Phys. Rev. Lett. **103**, 210401 (2009)
16. C.W. Helstrom, *Quantum Detection and Estimation Theory* (Academic, New York, 1976)
17. M.B. Ruskai, Beyond strong subadditivity? Improved bounds on the contraction of generalized relative entropy. Rev. Math. Phys. **6**, 1147 (1994)
18. B.-H. Liu, L. Li, Y.-F. Huang, C.-F. Li, G.-C. Guo, E.-M. Laine, H.-P. Breuer, J. Piilo, Experimental control of the transition from Markovian to non-Markovian dynamics of open quantum systems. Nat. Phys. **7**, 931 (2011)
19. E.-M. Laine, J. Piilo, H.-P. Breuer, Witness for initial system-environment correlations in open system dynamics. EPL **92**, 60010 (2010)
20. A. Rivas, S.F. Huelga, *Open Quantum Systems: An Introduction* (Springer, Berlin, 2012)
21. A. Rivas, S.F. Huelga, M.B. Plenio, Entanglement and non-Markovianity of quantum evolutions. Phys. Rev. Lett. **105**, 050403 (2010)
22. S. Wißmann, A. Karlsson, E.-M. Laine, J. Piilo, H.-P. Breuer, Optimal state pairs for non-Markovian quantum dynamics. Phys. Rev. A **86**, 062108 (2012)
23. S. Wißmann, H.-P. Breuer, B. Vacchini, Generalized trace-distance measure connecting quantum and classical non-Markovianity. Phys. Rev. A **92**, 042108 (2015)
24. H.-P. Breuer, G. Amato, B. Vacchini, Mixing-induced quantum non-Markovianity and information flow. New J. Phys. **20**, 043007 (2018)
25. D. Chruscinski, A. Kossakowski, A. Rivas, On measures of non-Markovianity: divisibility vs. backflow of information. Phys. Rev. A **83**, 052128 (2011)
26. J. Liu, X.-M. Lu, X. Wang, Nonunital non-Markovianity of quantum dynamics. Phys. Rev. A **87**, 042103 (2013)
27. A. Kossakowski, Necessary and sufficient conditions for a generator of a quantum dynamical semigroup. Bull. Acad. Pol. Sci. Ser. Math. Astr. Phys. **20**, 1021 (1972)
28. G. Amato, H.-P. Breuer, B. Vacchini, Generalized trace distance approach to quantum non-Markovianity and detection of initial correlations. Phys. Rev. A **98**, 012120 (2018)
29. A.A. Budini, Stochastic representation of a class of non-Markovian completely positive evolutions. Phys. Rev. A **69**, 042107 (2004)
30. D. Chruscinski, On time-local generators of quantum evolution. Open Syst. Inf. Dyn. **21**, 1440004 (2014)

31. A. Kossakowski, R. Rebolledo, On the structure of generators for non-Markovian master equations. Open Syst. Inf. Dyn. **16**, 259 (2009)
32. B. Vacchini, Non-Markovian master equations from piecewise dynamics. Phys. Rev. A **87**, 030101(R) (2013)
33. B. Vacchini, General structure of quantum collisional models. Int. J. Quantum Inform. **12**, 1461011 (2014)
34. D. Chruscinski, A. Kossakowski, Sufficient conditions for a memory-kernel master equation. Phys. Rev. A **94**, 020103(R) (2016)
35. B. Vacchini, Generalized master equations leading to completely positive dynamics. Phys. Rev. Lett. **117**, 230401 (2016)
36. D. Chruscinski, A. Kossakowski, Generalized semi-Markov quantum evolution. Phys. Rev. A **95**, 042131 (2017)
37. V. Nollau, *Semi-Markovsche Prozesse* (Akademie-Verlag, Berlin, 1980)
38. A. Barchielli, M. Gregoratti, *Quantum Trajectories and Measurements in Continuous Time.* Lecture Notes in Physics, vol. 782 (Springer, Berlin, 2009)
39. A.S. Holevo, *Statistical Structure of Quantum Theory.* Lecture Notes in Physics, vol. 67 (Springer, Berlin, 2001)
40. H.-P. Breuer, B. Vacchini, Quantum semi-Markov processes. Phys. Rev. Lett. **101**, 140402 (2008)
41. H.-P. Breuer, B. Vacchini, Structure of completely positive quantum master equations with memory kernel. Phys. Rev. E **79**, 041147 (2009)
42. W. Feller, On semi-Markov processes. Proc. Natl. Acad. Sci. **51**, 653 (1964)
43. G. Raithel, C. Wagner, H. Walther, L.M. Narducci, M.O. Scully, in *Cavity Quantum Electrodynamics*, ed. by P.R. Berman (Academic, San Diego, 1994), pp. 57–121
44. J. D. Cresser, Quantum-field model of the injected atomic beam in the micromaser. Phys. Rev. A **46**, 5913 (1992)
45. J.D. Cresser, S.M. Pickles, A quantum trajectory analysis of the one-atom micromaser. J. Opt. B Quantum Semiclass. Opt. **8**, 73 (1996)
46. S. Lorenzo, F. Ciccarello, G.M. Palma, Class of exact memory-kernel master equations. Phys. Rev. A **93**, 052111 (2016)
47. S. Lorenzo, F. Ciccarello, G.M. Palma, B. Vacchini, Quantum non-Markovian piecewise dynamics from collision models. Open Syst. Inf. Dyn. **24**, 1740011 (2017)

Chapter 5
Geometric Constructions over $\mathbb{C}$ and $\mathbb{F}_2$ for Quantum Information

Frédéric Holweck

Abstract In this review paper I present two geometric constructions of distinguished nature, one is over the field of complex numbers $\mathbb{C}$ and the other one is over the two elements field $\mathbb{F}_2$. Both constructions have been employed in the past 15 years to describe two quantum paradoxes or two resources of quantum information: entanglement of pure multipartite systems on one side and contextuality on the other. Both geometric constructions are linked to representation of semi-simple Lie groups/algebras. To emphasize this aspect one explains on one hand how well-known results in representation theory allows one to see all the classification of entanglement classes of various tripartite quantum systems (three qubits, three fermions, three bosonic qubits...) in a unified picture. On the other hand, one also shows how some weight diagrams of simple Lie groups are encapsulated in the geometry which deals with the commutation relations of the generalized N-Pauli group.

5.1 Introduction

The aim of this paper is to provide an elementary introduction to a series of papers involving geometrical descriptions of two different problems in quantum information theory: the classification of entanglement classes for pure multipartite quantum systems on one hand [32, 33, 35–37, 40, 51] and the observable-based proofs of the Kochen-Specker Theorem on the other hand [34, 38, 56, 67]. Apparently, these two problems have no direct connections to each other and the geometrical constructions to describe them are of distinguished nature. We will use projective complex geometry to describe entanglement classes and we will work with finite geometry over the two elements field $\mathbb{F}_2$ to describe operator-based proofs of the

F. Holweck (✉)
Laboratoire Interdisciplinaire Carnot de Bourgogne, University Bourgogne Franche-Comté, Belfort, France
e-mail: frederic.holweck@utbm.fr

© Springer Nature Switzerland AG 2019
E. Ballico et al. (eds.), *Quantum Physics and Geometry*,
Lecture Notes of the Unione Matematica Italiana 25,
https://doi.org/10.1007/978-3-030-06122-7_5

Kochen-Specker Theorem. However, when we look at both geometries from a representation theory point of view, one observes that the same semi-simple Lie groups are acting behind the scene. This observation may invite us to look for a more direct (physical) connection between those two questions. In this presentation I will also try to give many references on related works. However, this will not be an exhaustive review on all possible links between geometry and quantum information and there will be some references missing.

Before going into the details of the geometry, let us recall how those two questions are historically related to the question of the existence of hidden variables theories.

In the history of the development of quantum science, the paradoxes raised by questioning the foundations of quantum physics turn out to be considered as quantum resources once they have been tested experimentally. The most famous example of such a change of status for a scientific question is, of course, the EPR paradox which started by a criticism of the foundation of quantum physics by Einstein Podolsky and Rosen [26].

The famous EPR paradox deals with what we nowadays call a pure two-qubit quantum system. This is a physical system made of two parts A and B such that each part, or each particle, is a two-level quantum system. Mathematically, a pure two-qubit state is a vector of $\mathcal{H}_{AB} = \mathbb{C}_A^2 \otimes \mathbb{C}_B^2$. Denote by $(|0\rangle, |1\rangle)$ the standard basis of the vector spaces $\mathbb{C}_A^2$ and $\mathbb{C}_B^2$ and let $(|00\rangle, |01\rangle, |10\rangle, |11\rangle)$ be the associated basis of $\mathcal{H}_{AB}$. The laws of quantum mechanics tell us that $|\psi\rangle \in \mathcal{H}_{AB}$ can be described as

$$|\psi\rangle = a_{00} |00\rangle + a_{10} |10\rangle + a_{01} |01\rangle + a_{11} |11\rangle, \tag{5.1}$$

with $a_{ij} \in \mathbb{C}$ and $|a_{00}|^2 + |a_{10}|^2 + |a_{01}|^2 + |a_{11}|^2 = 1$. Einstein, Podolsky and Rosen introduced the following admissible state

$$|EPR\rangle = \frac{1}{\sqrt{2}}(|00\rangle + |11\rangle), \tag{5.2}$$

to argue that quantum mechanics was incomplete. The EPR reasoning consists of saying that, according to quantum mechanics, a measurement of particle A will project the system $|EPR\rangle$ to either $|00\rangle$ or $|11\rangle$, fixing instantaneously the possible outcomes of the measurement of particle B no matter how far the distance between particles A and B is. This was characterized in [26] as *spooky action at the distance* and according to Einstein, Podolsky and Rosen this was showing that *hidden variables* were necessary to make the theory complete. Note that none of all two-qubit quantum states can produce a spooky action at the distance. If $|\psi\rangle = (\alpha_A |0\rangle + \beta_A |1\rangle) \otimes (\alpha_B |0\rangle + \beta_B |1\rangle)$, then the measurement of particle A has no impact on the state of particle B. From Eq. (5.1) one sees that the possibility to factorize a state $|\psi\rangle$ translates to

$$a_{00}a_{11} - a_{01}a_{10} = 0. \tag{5.3}$$

Fig. 5.1 In
$\mathbb{P}^3 = \mathbb{P}(\mathbb{C}^2 \otimes \mathbb{C}^2)$, the zero
set of the quadric,
$a_{00}a_{11} - a_{01}a_{10} = 0$, is
denoted by X_{Sep} and
corresponds to the
non-entangled (or separable)
states. The entangled states
are defined by the
complement, $\mathbb{P}^3 \backslash X_{\text{Sep}}$

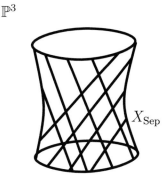

This homogeneous equation defines a quadratic hypersurface in $\mathbb{P}^3 = \mathbb{P}(\mathbb{C}^2 \otimes \mathbb{C}^2)$, corresponding to the projectivization of the states that can be factorized; those states are called *non-entangled states*. The complement of the quadric is the set of non-factorizable states, i.e. *entangled states* (Fig. 5.1).

The philosophical questioning of Einstein and his co-authors about the existence of hidden-variables to make quantum physics complete becomes a scientific question after the work of John Bell [9], 30 years later, whose inequalities have opened up the path to experimental tests. Those experimental tests have been performed many times starting with the pioneering works of Alain Aspect [5] and *entanglement* in multipartite systems is nowadays recognized as an essential resource in quantum information.

Another paradox of quantum physics, maybe less famous than EPR, is *contextuality*. Interestingly, the notion of contextuality in quantum physics is also related to the question of the existence of hidden-variables. In 1975 Kochen and Specker[1] [44] introduced this notion by proving there is no non-contextual hidden-variables theory which can reproduce the outcomes predicted by quantum physics. Here *contextual* means that the outcome of a measurement on a quantum system depends on the context, i.e. a set of compatible measurements (set of mutually commuting observables[2]) that are performed in the same experiment. The original proof of Kochen and Specker is based on the impossibility to assign coloring (i.e. predefine values for the outcomes) to some vector basis associated to some set of projection operators. Let us present here a simple and nice observable-based proof of the

---

[1]This concept of contextuality also appears in Bell's paper [9, 59].

[2]In quantum physics, the outcomes of a measurement are encoded in an hermitian operator, called an observable. The eigenvalues of the observable correspond to the possible outcomes of the measurement and the eigenvectors correspond to the possible projections of the state after measurement.

Fig. 5.2 The Mermin-Peres "Magic" square: Each node is an element of the two-qubit Pauli group which squares to I_4. Each row and column represents a set of compatible observables (mutually commuting operators) such that their product equals $\pm I_4$

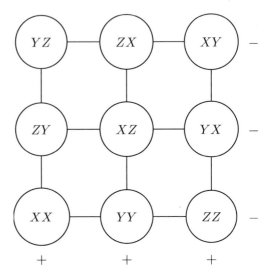

Kochen-Specker Theorem due to Mermin [59] and Peres [68]. Let us denote by X, Y and Z the usual Pauli matrices,

$$X = \begin{pmatrix} 0 & 1 \\ 1 & 0 \end{pmatrix}, \quad Y = \begin{pmatrix} 0 & -i \\ i & 0 \end{pmatrix}, \quad Z = \begin{pmatrix} 1 & 0 \\ 0 & -1 \end{pmatrix}. \tag{5.4}$$

Those three hermitian operators encode the possible measurement outcomes of a spin-$\frac{1}{2}$-particle in a Stern-Gerlach apparatus oriented in three different space directions. Taking tensor products of two such Pauli matrices we can define Pauli operators acting on two qubits. In [59, 68] Mermin and Peres considered a set of two-qubit Pauli operators similar to the one reproduced in Fig. 5.2.

This diagram, called the "Magic" Mermin-Peres square, furnishes a proof of the impossibility to predict the outcomes of quantum physics with a non-contextual hidden-variables theory as I now explain. Each node of the square represents a two-qubit observable which squares to identity, i.e. the possible eigenvalues of each node (the possible measurement outcomes) are ± 1. The operators which belong to a row or a column are mutually commuting, i.e. they represent a context or a set of compatible observables. The products of each row or column give either I_4 or $-I_4$ as indicated by the signs on the diagram. The odd number of negative rows makes it impossible to pre-assign to each node outcomes (± 1) which are simultaneously compatible with the constrains on the rows (the products of the eigenvalues should be negative) and columns (the product of the eigenvalues are positive). Therefore, any hidden-variables theory capable of reproducing the outcomes of the measurement that can be achieved with the Mermin-Peres square, should be contextual, i.e. the deterministic values that we wish to assign should be context dependent. This other paradox has been studied intensively in the last decade and experiments [1, 8, 18, 43] are now conducted to produce contextuality in the laboratory, leading to consider contextuality as another quantum resource for quantum computation or quantum processing [1, 41].

Both entanglement of multipartite pure quantum systems and contextual configurations of multi-qubit Pauli observables can be nicely described by geometric constructions. To put into perspective those two problems and their corresponding geometric descriptions, I choose to emphasize their relations with representation theory. In Sect. 5.2, I introduce the geometric language of auxiliary varieties and I explain how various classification results introduced in the quantum information literature in the past 15 years can be uniformly described in terms of representation theory. In Sect. 5.3, I describe geometrically the set of commutation relations within the N-qubit Pauli group and explain through explicit examples how weight diagrams of some simple Lie algebras can be extracted from such commutation relations.

5.2   The Geometry of Entanglement

In the first part of the paper, I discuss the question of the classification of entanglement for multipartite quantum systems under the SLOCC group from the point of view of algebraic geometry and representation theory. In the past 15 years, there have been a lot of papers on the subject tackling the classification for different types of quantum systems [11, 12, 16, 20, 22, 24, 25, 50, 53, 60, 61, 74]. The most famous one is probably the paper of Dür, Vidal and Cirac [25] where it was shown that three-qubit quantum states can be genuinely entangled in two different ways.

5.2.1   Entanglement Under SLOCC, Tensor Rank and Algebraic Geometry

The Hilbert space of an n-partite system will be the tensor product of n-vector spaces, where each vector space is the Hilbert space of each individual part. Thus the Hilbert space of an n-qudit system is $\mathcal{H} = \mathbb{C}^{d_1} \otimes \cdots \otimes \mathbb{C}^{d_n}$. A quantum state being defined up to a phase, we will work in the projective Hilbert space and denote by $[\psi] \in \mathbb{P}(\mathcal{H})$ the class of a quantum state $|\psi\rangle \in \mathcal{H}$. The group of local reversible operations, $G = \mathrm{SL}_{d_1}(\mathbb{C}) \times \cdots \times \mathrm{SL}_{d_n}(\mathbb{C})$ acts on $\mathbb{P}(\mathcal{H})$ by its natural action. This group is known in physics as the group of Stochastic Local Operations with Classical Communications [10, 25] and will be denoted by SLOCC.

According to the axioms of quantum physics, it would be more natural to look at entanglement classes of multipartite quantum systems under the group of Local Unitary transformations, LU= $\mathrm{SU}(d_1) \times \cdots \times \mathrm{SU}(d_n)$. In quantum information theory one also considers a larger set of transformations called LOCC transformations (Local Operations with Classical Communications) which include local unitary and measurement operations (coordinated by classical communications).

Under LOCC two quantum states are equivalent if they can be exactly interconverted by LU operations.[3] However, the SLOCC equivalence also has a physical meaning as explained in [10, 25]. It corresponds to an equivalence between states that can be interconverted into each other but not with certainty. Another feature of SLOCC is that if we consider measure of entanglement, the amount of entanglement may increase or decrease under SLOCC while it is invariant under LU and non-increasing under LOCC. However, entanglement cannot be created or destroyed by SLOCC and a communication protocol based on a quantum state $|\psi_1\rangle$ can also be achieved with a SLOCC equivalent state $|\psi_2\rangle$ (eventually with different probability of success). In this sense, SLOCC equivalence is more a qualitative way of separating non equivalent quantum states.

The set of separable, or non-entangled states, is the set of quantum states $|\psi\rangle$ which can be factorized, i.e.

$$|\psi\rangle = |\psi_1\rangle \otimes \cdots \otimes |\psi_n\rangle \text{ with } |\psi_k\rangle \in \mathbb{C}^{d_k}. \tag{5.5}$$

In algebraic geometry the projectivization of this set is a well-known algebraic variety[4] of $\mathbb{P}(\mathcal{H})$, known as the Segre embedding of the product of projective spaces $\mathbb{P}^{d_1-1} \times \cdots \times \mathbb{P}^{d_n-1}$.

More precisely, let us consider the following map,

$$\begin{aligned} Seg : \mathbb{P}^{d_1-1} \times \cdots \times \mathbb{P}^{d_n-1} &\to \mathbb{P}^{d_1\times\cdots\times d_n-1} = \mathbb{P}(\mathcal{H}) \\ ([\psi_1], \ldots, [\psi_n]) &\mapsto [\psi_1 \otimes \cdots \otimes \psi_n]. \end{aligned} \tag{5.6}$$

The image of this map is the Segre embedding of the product of projective spaces and clearly coincides with X_{Sep}, the projectivization of the set of separable states. We will thus write

$$X_{\text{Sep}} = \mathbb{P}^{d_1-1} \times \cdots \times \mathbb{P}^{d_n-1} \subset \mathbb{P}(\mathcal{H}). \tag{5.7}$$

The Segre variety has the property to be the only one closed orbit of $\mathbb{P}(\mathcal{H})$ for the SLOCC action. Up to local reversible transformations, every separable state $|\psi\rangle = |\psi_1\rangle\otimes\cdots\otimes|\psi_n\rangle$ can be transformed to $|0\rangle\otimes\cdots\otimes|0\rangle = |0\ldots0\rangle$ if we assume that each vector space $\mathbb{C}^{d_i}$ is equipped with a basis denoted by $|0\rangle, \ldots, |d_i - 1\rangle$,

$$X_{\text{Sep}} = \mathbb{P}^{d_1-1} \times \cdots \times \mathbb{P}^{d_n-1} = \mathbb{P}(\text{SLOCC}.|0\ldots0\rangle) \subset \mathbb{P}(\mathcal{H}). \tag{5.8}$$

---

[3]Physically one may imagine that each part of the system is in a different location and experimentalists only apply local quantum transformations, i.e. some unitaries defined by local Hamiltonians.

[4]In this paper an algebraic variety will always be the zero locus of a collection of homogeneous polynomials.

A quantum state $|\psi\rangle \in \mathcal{H}$ is entangled iff it is not separable, i.e.

$$|\psi\rangle \text{ entangled} \iff [\psi] \in \mathbb{P}(\mathcal{H} \setminus X_{\text{Sep}}). \tag{5.9}$$

In algebraic geometry, it is usual to study properties of X by introducing auxiliary varieties, i.e. varieties built from the knowledge of X, whose attributes (dimension, degree) will tell us something about the geometry of X.

Let us first introduce two auxiliary varieties of importance for quantum information and entanglement: the secant and tangential varieties.

Definition 5.1 Let $X \subset \mathbb{P}(V)$ be a projective algebraic variety, the secant variety of X is the Zariski closure of the union of secant lines, i.e.

$$\sigma_2(X) = \overline{\cup_{x,y \in X} \mathbb{P}^1_{xy}}, \tag{5.10}$$

where $\mathbb{P}^1_{xy}$ is the projective line corresponding to the projectivization of the linear span $\text{Span}(\hat{x}, \hat{y}) \subset V$ (a 2-dimensional linear subspace of V).

Remark 5.1 This definition can be extended to higher-dimensional secant varieties. More generally, one may define the kth-secant variety of X,

$$\sigma_k(X) = \overline{\cup_{x_1,\dots,x_k} \mathbb{P}^{k-1}_{x_1,\dots,x_k}}, \tag{5.11}$$

where now $\mathbb{P}^{k-1}_{x_1,\dots,x_k}$ is the a projective subspace of dimension $k-1$ obtained as the projectivization of the linear span $\text{Span}(\hat{x}_1, \dots, \hat{x}_n) \subset V$. There is a natural sequence of inclusions given by $X \subset \sigma_2(X) \subset \sigma_3(X) \subset \cdots \subset \sigma_q(X) = \mathbb{P}(V)$, where q is the smallest integer such that the qth-secant variety fills the ambient space.

Remark 5.2 The notion of secant varieties is deeply connected to the notion of rank of tensors. One says that a tensor $T \in \mathbb{C}^{d_1} \otimes \cdots \otimes \mathbb{C}^{d_n}$ has rank r iff r is the smallest integer such that $T = T_1 + \cdots + T_r$ and each tensor T_i can be factorized, i.e. $T_i = a_1^i \otimes \cdots \otimes a_n^i$. From the definition one sees that the Segre variety $\mathbb{P}^{d_1-1} \times \cdots \times \mathbb{P}^{d_n-1}$ corresponds to the projectivization of rank-one tensors of $\mathcal{H}$ and the secant variety of the Segre is the Zariski closure of the (projectivization of) rank-two tensors because a generic point of $\sigma_2(\mathbb{P}^{d_1-1} \times \cdots \times \mathbb{P}^{d_n-1})$ is the sum of two rank-one tensors. Similarly, $\sigma_k(\mathbb{P}^{d_1-1} \times \cdots \times \mathbb{P}^{d_n-1})$ is the algebraic closure of the set of rank at most k tensors. Tensors (states) which belong to $\sigma_k(\mathbb{P}^{d_1-1} \times \cdots \times \mathbb{P}^{d_n-1}) \setminus \sigma_{k-1}(\mathbb{P}^{d_1-1} \times \cdots \times \mathbb{P}^{d_n-1})$ will be called tensors (states) of border rank-k, i.e. they can be expressed as (limits) of rank k tensors.

Another auxiliary variety of importance is the tangential variety, i.e. the union of tangent spaces. When $x \in X$ is a smooth point of the variety I denote by $T_x X$ the projective tangent space and $\hat{T}_x X$ its cone in $\mathcal{H}$.

Definition 5.2 Let $X \subset \mathbb{P}(V)$ be a smooth projective algebraic variety, the tangential variety of X is defined by

$$\tau(X) = \cup_{x \in X} T_x X, \tag{5.12}$$

(here the smoothness of X implies that the union is closed).

The auxiliary varieties built from X_{Sep} are of importance to understand the entanglement stratification of Hilbert spaces of pure quantum systems under SLOCC for mainly two reasons. First the auxiliary varieties are SLOCC invariants by construction because X_{Sep} is a SLOCC-orbit. Thus the construction of auxiliary varieties from the core set of separable states X_{Sep} produces a stratification of the ambient space by SLOCC-invariant algebraic varieties. The possibility to stratify the ambient space by secant varieties was known to geometers more than a century ago [76], but it was noticed to be useful for studying entanglement classes only recently by Heydari [31]. It is equivalent to a stratification of the ambient space by the (border) ranks of the states which, as pointed out by Brylinski, can be considered as an algebraic measure of entanglement [17].

The second interesting aspect of those auxiliary varieties, in particular the secant and tangent one, is that they may have a nice quantum information interpretation. To be more precise, let us recall the definition of the $|GHZ_n\rangle$ and $|W_n\rangle$ states,

$$|GHZ_n\rangle = \frac{1}{\sqrt{2}}(|0\ldots0\rangle + |1\ldots1\rangle), \tag{5.13}$$

$$|W_n\rangle = \frac{1}{\sqrt{n}}(|100\ldots0\rangle + |010\ldots0\rangle + \cdots + |00\ldots1\rangle). \tag{5.14}$$

Then we have the following geometric interpretations of the closure of their corresponding SLOCC classes,

$$\overline{\text{SLOCC.}[GHZ_n]} = \sigma_2(X_{Sep}) \text{ and } \overline{\text{SLOCC.}[W_n]} = \tau(X_{Sep}). \tag{5.15}$$

It is not difficult to see why the Zariski closure of the SLOCC orbit of the $|GHZ_n\rangle$ state is the secant variety of the set of separable states. Recall that a generic point of $\sigma(X_{Sep})$ is a rank 2 tensor. Thus, if $[z]$ is a generic point of $\sigma_2(X_{Sep})$, one has

$$[z] = [\lambda x_1 \otimes x_2 \otimes \cdots \otimes x_n + \mu y_1 \otimes y_2 \otimes \cdots \otimes y_n], \tag{5.16}$$

with $x_i, y_i \in \mathbb{C}^{d_i}$. Because $[z]$ is generic we may assume that (x_i, y_i) are linearly independent. Therefore, there exists $g_i \in \text{SL}_{d_i}(\mathbb{C})$ such that $g_i.x_i \propto |0\rangle$ and $g_i.y_i \propto |1\rangle$ for all $i \in \{1, \ldots, n\}$. Thus we can always find $g \in \text{SLOCC}$ such that $[g.z] = [GHZ_n]$.

To see why the tangential variety of the variety of separable states always corresponds to the (projective) orbit closure of the $|W_n\rangle$ state, we need to show that a generic tangent vector of X_{Sep} is always SLOCC equivalent to $|W_n\rangle$. A tangent vector can be obtained by differentiating a curve of X_{Sep}. Let $[x(t)] = [x_1(t) \otimes x_2(t) \otimes \cdots \otimes x_n(t)] \subset X_{\text{Sep}}$ with $[x(0)] = [x_1 \otimes x_2 \otimes \cdots \otimes x_n]$. Because we are looking at a generic tangent vector, we assume that for all i, $x_i'(0) = u_i$ and u_i is not collinear to x_i. Then Leibniz's rule insures that

$$[x'(0)] = [u_1 \otimes x_2 \otimes \cdots \otimes x_n + x_1 \otimes u_2 \otimes \cdots \otimes x_n + \cdots + x_1 \otimes x_2 \otimes \cdots \otimes u_n]. \quad (5.17)$$

Let us consider $g_i \in \mathrm{SL}_{d_i}(\mathbb{C})$ such that $g_i.x_i \propto |0\rangle$ and $g_i.u_i \propto |1\rangle$, then we obtain $[g.x'(0)] = [W_n]$ for $g = (g_1, \ldots, g_n)$.

An important result regarding the relationship between tangent and secant varieties is due to Fyodor Zak [84].

Theorem 5.1 ([84]) *Let $X \subset \mathbb{P}(V)$ be a projective algebraic variety of dimension d. Then one of the following two properties holds,*

1. *$dim(\sigma_2(X)) = 2d + 1$ and $dim(\tau(X)) = 2d$,*
2. *$dim(\sigma_2(X)) \leq 2d$ and $\tau(X) = \sigma_2(X)$.*

To get information from Zak's theorem one needs to compute the dimension of the secant variety of X. This can be done by using an old geometrical result from the beginning of the twentieth century known as Terracini's Lemma.

Lemma 5.1 (Terracini's Lemma) *Let $[z] \in \sigma_2(X)$ with $[z] = [x + y]$ and $([x], [y]) \in X \times X$ be a general pair of points. Then*

$$\hat{T}_{[z]}\sigma_2(X) = \hat{T}_{[x]}X + \hat{T}_{[y]}X. \quad (5.18)$$

Terracini's Lemma tells us that if X is of dimension d, the expected dimension of $\sigma_2(X)$ is $2(d + 1) - 1 = 2d + 1$. Thus by Zak's Theorem, one knows that if $\sigma_2(X)$ has the expected dimension then the tangential variety is a proper subvariety of $\sigma_2(X)$ and otherwise both varieties are the same.

Example Let us look at the case where $X_{\text{Sep}} = \mathbb{P}^1 \times \mathbb{P}^1 \times \mathbb{P}^1 \subset \mathbb{P}^7$. The dimension of $\sigma_2(X_{\text{Sep}})$ can be obtained as a simple application of Terracini's lemma. Let $[x] = [\phi_1 \otimes \phi_2 \otimes \phi_3] \in X_{\text{Sep}}$ then $\hat{T}_{[x]}X_{\text{Sep}} = \mathbb{C}^2 \otimes \phi_2 \otimes \phi_3 + \phi_1 \otimes \mathbb{C}^2 \otimes \phi_3 + \phi_1 \otimes \phi_2 \otimes \mathbb{C}^2$. Thus one gets for $[GHZ] = [|000\rangle + |111\rangle] \in \sigma_2(X_{\text{Sep}})$,

$$\begin{aligned}
\hat{T}_{[GHZ]}\sigma_2(X_{\text{Sep}}) &= \hat{T}_{[|000\rangle]}X_{\text{Sep}} + \hat{T}_{[|111\rangle]}X_{\text{Sep}} \\
&= \mathbb{C}^2 \otimes |0\rangle \otimes |0\rangle + |0\rangle \otimes \mathbb{C}^2 \otimes |0\rangle + |0\rangle \otimes |0\rangle \otimes \mathbb{C}^2 \\
&\quad + \mathbb{C}^2 \otimes |1\rangle \otimes |1\rangle + |1\rangle \otimes \mathbb{C}^2 \otimes |1\rangle + |1\rangle \otimes |1\rangle \otimes \mathbb{C}^2.
\end{aligned}$$
$$(5.19)$$

Therefore $\dim(\hat{T}_{[GHZ]}\sigma_2(X_{\text{Sep}})) = 8$, i.e. $\dim(\sigma_2(X_{\text{Sep}})) = 7$.

5.2.2   The Three-Qubit Classification via Auxiliary Varieties

As mentioned at the beginning of the section, the problem of the classification of multipartite quantum systems acquired a lot of attention after Dür, Vidal and Cirac's paper [25] on the classification of three-qubit states, where it was first shown that two quantum states can be entangled in two genuine non-equivalent ways. The authors showed that for three-qubit systems there are exactly six SLOCC orbits whose representatives can be chosen to be: $|Sep\rangle = |000\rangle$, $|B_1\rangle = \frac{1}{\sqrt{2}}(|000\rangle + |011\rangle)$, $|B_2\rangle = \frac{1}{\sqrt{2}}(|000\rangle + |101\rangle)$, $|B_3\rangle = \frac{1}{\sqrt{2}}(|000\rangle + |110\rangle)$, $|W_3\rangle = \frac{1}{\sqrt{3}}(|100\rangle + |010\rangle + |001\rangle)$ and $|GHZ_3\rangle = \frac{1}{\sqrt{2}}(|000\rangle + |111\rangle)$.

The state $|Sep\rangle$ is a representative of the orbit of separable states and the states $|B_i\rangle$ are bi-separable. The only genuinely entangled states are $|W_3\rangle$ and $|GHZ_3\rangle$. It turns out that this orbit classification of the Hilbert space of three qubits was known long before the famous paper of Dür, Vidal and Cirac from different mathematical perspectives (see for example [28, 65]). Probably the oldest mathematical proof of this result goes back to the work of Le Paige (1881) who classified the trilinear binary forms under (local) linear transformations in [49].

From a geometrical point of view the existence of two distinguished orbits corresponding to $|W_3\rangle$ and $|GHZ_3\rangle$ can be obtained as a consequence of Zak's theorem (Theorem 5.1). Indeed, one shows, Eq. (5.19), that the secant variety of the variety of separable three qubit states has the expected dimension and fills the ambient space. According to Zak's Theorem, this implies that the tangential variety $\tau(X_{Sep})$ is a codimension-one sub-variety of $\sigma_2(X_{Sep}) = \mathbb{P}^7$ and, therefore, both orbits are distinguished. In other words, from a geometrical perspective there exist two non-equivalent, genuinely entangled states for the three-qubit system because the secant variety of the set of separable states has the expected dimension and fills the ambient space.

In this language of auxiliary varieties let us also mention that the orbit closures defined by the bi-separable states $|B_i\rangle$ have also a geometric interpretation. For instance, $|B_1\rangle = |0\rangle \otimes \frac{1}{\sqrt{2}}(|00\rangle + |11\rangle) = |0\rangle \otimes |EPR\rangle$. The projective orbit closure is

$$\mathbb{P}(\overline{SLOCC. |B_1\rangle}) = \mathbb{P}^1 \times \mathbb{P}^3 \subset \mathbb{P}^7, \qquad (5.20)$$

where $\mathbb{P}^3 = \sigma_2(\mathbb{P}^1 \times \mathbb{P}^1)$. The geometric stratification by SLOCC invariant algebraic varieties in the three-qubit case can be represented as in Fig. 5.3.

Remark 5.3 This idea of introducing auxiliary varieties to describe SLOCC classes of entanglement also appears in [73, 75]. It allows one to connect the study of entanglement in quantum information to a large literature in mathematics, geometry and their applications. For instance, the question of finding defining equations of auxiliary varieties is central in many areas of applications of mathematics to computer science, signal processing or phylogenetics (see the introduction of [45]

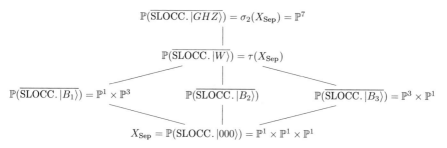

Fig. 5.3 Stratification of the (projectivized) Hilbert space of three qubits by SLOCC-invariant algebraic varieties (the secant and tangent)

and references therein). Those equations can be obtained by mixing techniques from representation theory and geometry [47, 48, 64]. In the context of quantum information finding defining equations of auxiliary varieties provides tests to decide if two states could be SLOCC equivalent. Classical invariant theory also provides tools to generate invariant and covariant polynomials [14, 15, 57, 58] and these techniques were used in [35, 36, 40] to identify entanglement classes with auxiliary varieties.

5.2.3   Geometry of Hyperplanes: The Dual Variety

Another auxiliary variety of interest is the dual variety of X_{Sep}:

$$X^*_{\mathrm{Sep}} = \overline{\{H \in (\mathbb{P}^N)^*, \exists x \in X_{\mathrm{Sep}}, T_x X_{\mathrm{Sep}} \subset H\}}. \tag{5.21}$$

The variety X^*_{Sep} parametrizes the set of hyperplanes defining singular (non-smooth) hyperplane sections of X_{Sep}. Using the hermitian inner product on $\mathcal{H}$, one can identify the dual variety of X_{Sep} with the set of states which define a singular hyperplane section of X_{Sep}. More precisely, given a state $|\psi\rangle \in \mathcal{H}$ we have

$$[\psi] \in X^*_{\mathrm{Sep}} \text{ iff } X_{\mathrm{Sep}} \cap H_\psi = \{[\varphi] \in X_{\mathrm{Sep}}, \langle \psi, \varphi \rangle = 0\} \text{ is singular.} \tag{5.22}$$

For $X_{\mathrm{Sep}} = \mathbb{P}^{d_1-1} \times \mathbb{P}^{d_2-1} \times \cdots \times \mathbb{P}^{d_n-1}$ (with $d_j \leq \sum_{i \neq j} d_i$), the variety X^*_{Sep} is always a hypersurface, called the hyperdeterminant of format $d_1 \times d_2 \times \cdots \times d_n$ [28]. By construction the hyperdeterminant is SLOCC-invariant and so is its singular locus. Therefore, the hyperdeterminant and its singular locus can be used to stratify the (projectivized) Hilbert space under SLOCC.

This idea goes back to Miyake [60–62] who interpreted previous results of Weyman and Zelevinsky on singularities of hyperdeterminants [83] to describe the entanglement structure for the three- and four-qubit systems, as well as for the

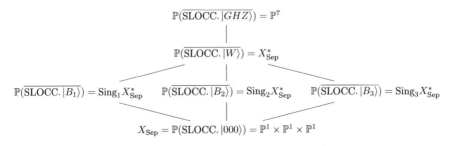

Fig. 5.4 Stratification of the (projectivized) Hilbert space of three qubit by SLOCC-invariant algebraic varieties (the dual and its singular locus). $\text{Sing}_i X^*_{\text{Sep}}$ represent different components of the singular locus [60, 83]

$2 \times 2 \times n$-systems. Following Miyake, the hyperdeterminant of format $2 \times 2 \times 2$, also known as the Cayley hyperdeterminant, provides a dual picture of the three-qubit classification (Fig. 5.4).

One can go further by studying which types of singular hyperplane sections can be associated to a given state.

To do so we use the rational map defining the Segre embedding to obtain the equations of the hyperplane sections:

$$\mathbb{P}^{d_1-1} \times \cdots \times \mathbb{P}^{d_n-1} \qquad \rightarrow \qquad \mathbb{P}(\mathcal{H})$$
$$([x_1^1 : \cdots : x_{d_1}^1], \ldots, [x_1^n : \cdots : x_{d_n}^n]) \mapsto [x_1^1 x_1^2 \ldots x_1^n : \cdots : \mathbf{x}_J : \cdots : x_{d_1}^1 x_{d_2}^2 \ldots x_{d_n}^n],$$
$$\tag{5.23}$$

where $\mathbf{x}_J$, for $J = (i_1, \ldots, i_n)$ with $1 \leq i_j \leq d_j$, denotes the monomial $\mathbf{x}_J = x_{i_1}^1 x_{i_2}^2 \ldots x_{i_n}^n$. In (5.23) the monomials $\mathbf{x}_J$ are ordered lexicographically in terms of multi-indices J. Therefore to a state $|\psi\rangle = \sum a_{i_1\ldots i_n} |i_1 \ldots i_n\rangle$ one associates the hypersurface of X_{Sep} defined by

$$f_{|\psi\rangle} = \sum_{i_1,\ldots,i_n} \overline{a_{i_1\ldots i_n}} x_{i_1}^1 \ldots x_{i_n}^n = 0. \tag{5.24}$$

If $|\psi\rangle \in X^*_{\text{Sep}}$, then $f_{|\psi\rangle}$ is a singular homogeneous polynomial, i.e. there exists $\tilde{x} \in X_{\text{Sep}}$ such that

$$f_{|\psi\rangle}(\tilde{x}) = 0 \text{ and } \partial_{i_k} f_{|\psi\rangle}(\tilde{x}) = 0. \tag{5.25}$$

In the 70s Arnol'd defined and classified *simple singularities* of complex functions [3, 4].

Definition 5.3 One says that $(f_{|\psi\rangle}, \tilde{x})$ is simple iff under a small perturbation it can only degenerate to a finite number of non-equivalent singular hypersurfaces $(f_{|\psi\rangle} + \varepsilon g, \tilde{x}')$ (up to biholomorphic change of coordinates).

Table 5.1 Simple singularities and their normal forms

Type	A_k	D_k	E_6	E_7	E_8
Normal form	$x^{k+1} + y^2$	$x^{k-1} + xy^2$	$x^3 + y^4$	$x^3 + xy^3$	$x^3 + y^5$
Milnor number	k	k	6	7	8

Simple singularities are always isolated, i.e. the Milnor number of the singularity $(f, \tilde{x})$, $\mu = \dim\mathbb{C}[x_1, \ldots, x_n]/(\nabla f_{\tilde{x}})$, is finite, and they can be classified in five families (Table 5.1).

The singular type can be identified by computing the Milnor number, the corank of the Hessian and the cubic term in the degenerate directions.

Example Let us consider the four-qubit state $|\psi\rangle = |0000\rangle + |1011\rangle + |1101\rangle + |1110\rangle$. The parametrization of the variety of separable states is given by $\phi([x_0 : x_1], [y_0 : y_1], [z_0 : z_1], [t_0 : t_1]) = [x_0 y_0 z_0 t_0 : \cdots : x_1 y_1 z_1 t_1]$. The homogeneous polynomial associated to $|\psi\rangle$ is

$$f_{|\psi\rangle} = x_0 y_0 z_0 t_0 + x_1 y_0 z_1 t_1 + x_1 y_1 z_0 t_1 + x_1 y_1 z_1 t_0. \tag{5.26}$$

In the chart $x_0 = y_1 = z_1 = t_1 = 1$ one obtains locally the hypersurface defined by

$$f(x, y, z, t) = yzt + xy + xz + xt. \tag{5.27}$$

The point $(0, 0, 0, 0)$ is the only singular point of $f_{|\psi\rangle}$ (the hyperplane section is tangent to $[|0111\rangle]$). The Hessian matrix of this singularity has co-rank 2 and $\mu = 4$. Therefore the hyperplane section defined by $|\psi\rangle$ has a unique singular point of type D_4 and this is true for all states SLOCC equivalent to $|\psi\rangle$.

The four-qubit and three-qutrit pure quantum systems are examples of systems with an infinite number of SLOCC-orbits. However, in both cases the orbit structure can still be described in terms of family of normal forms by introducing parameters. The four-qubit classification was originally obtained by Verstraete et al. [78] with a small correction provided by Chterental and Djoković [22]. Regarding the 3-qutrit classification, it has not been published in the quantum physics literature, but it can be directly translated from the orbit classification of the $3 \times 3 \times 3$ complex hypermatrices under $GL_3(\mathbb{C}) \times GL_3(\mathbb{C}) \times GL_3(\mathbb{C})$ obtained by Nurmiev [63]. In [32, 37] I calculated with my co-authors the type of isolated singularities associated to those forms. First of all, all isolated singularities are simple but moreover the worst, in terms of degeneracy, isolated singularity that arises is, in both cases, of type D_4. This allows us to get a more precise onion-like description [60] of the classification, see Fig. 5.5. It also gives information about how a state can be perturbed to another one. For instance, for a sufficiently small perturbation a state corresponding to a singular hyperplane section with only isolated singularities can only be changed to a state with isolated singularities of a lower degeneracy.

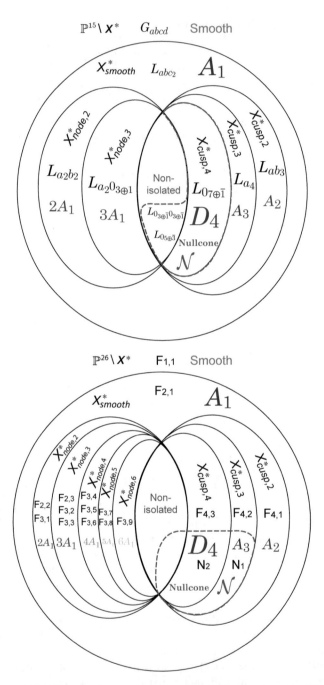

Fig. 5.5 Four-qubit and three-qutrit entanglement stratification by singular types of the hyperplane sections. Thus cusp components correspond to states with singularities which are not of type A_1 and the node components correspond to states with at least two singular points [83]. The names of the normal forms come from [78] and [63]

Theorem 5.2 ([37])   *Let H_ψ be a hyperplane of $\mathbb{P}(\mathcal{H})$ tangent to $X_{Sep} = \mathbb{P}^1 \times \mathbb{P}^1 \times \mathbb{P}^1 \times \mathbb{P}^1 \subset \mathbb{P}^{15}$ and such that $X_{Sep} \cap H_\psi$ has only isolated singular points. Then the singularities are either of types A_1, A_2, A_3, or of type D_4, and there exist hyperplanes realizing each type of singularity. Moreover, if we denote by $\widehat{X}^*_{Sep} \subset \mathcal{H}$ the cone over the dual variety of X_{Sep}, i.e. the zero locus of the Cayley hyperdeterminant of format $2 \times 2 \times 2 \times 2$, then the quotient map[5] $\Phi : \mathcal{H} \to \mathbb{C}^4$ is such that $\Phi(\widehat{X}^*_{Sep}) = \Sigma_{D_4}$, where Σ_{D_4} is the discriminant of the miniversal deformation[6] of the D_4-singularity.*

Theorem 5.3 ([32])   *Let $H_\psi \cap X$ be a singular hyperplane section of the algebraic variety of separable states for three-qutrit systems, i.e. $X_{Sep} = \mathbb{P}^2 \times \mathbb{P}^2 \times \mathbb{P}^2 \subset \mathbb{P}^{26}$ defined by a quantum pure state $[\psi] \in \mathbb{P}^{26}$. Then $H_\psi \cap X_{Sep}$ only admits simple or nonisolated singularities. Moreover if x is an isolated singular point of $H_\psi \cap X_{Sep}$, then its singular type is either A_1, A_2, A_3 or D_4.*

5.2.4   Representation Theory and Quantum Systems

Let us now consider G, a complex semi-simple Lie group, and V, an irreducible representation of G, i.e. one considers a map $\rho : G \to GL(V)$ defining an action of G on V such that there is no proper subspace of V stablized by G. The projectivization of an irreducible representation $\mathbb{P}(V)$ always contains a unique closed orbit $X_G \subset \mathbb{P}(V)$ called the highest weight orbit [27]. The Hilbert space $\mathcal{H} = \mathbb{C}^{d_1} \otimes \cdots \otimes \mathbb{C}^{d_n}$ is an irreducible representation of $SLOCC = SL_{d_1}(\mathbb{C}) \times \cdots \times SL_{d_n}(\mathbb{C})$ and, in this particular case, the highest weight orbit is nothing but the Segre variety $X_{Sep} = \mathbb{P}^{d_1-1} \times \cdots \times \mathbb{P}^{d_n-1}$.

It is natural to ask if other semi-simple Lie groups and representations have physical interpretations in terms of quantum systems. Let us first introduce the case of symmetric and skew-symmetric states.

- Consider the simple complex Lie group $SLOCC = SL_n(\mathbb{C})$ and its irreducible representation $\mathcal{H}_{bosons} = Sym^k(\mathbb{C}^n)$ where $Sym^k(\mathbb{C}^n)$ is the kth symmetric tensor product of $\mathbb{C}^n$. Then $\mathcal{H}_{bosons}$ is the Hilbert space of k indistinguishable symmetric particles, each particle being an n-single particle state. Physically, it

---

[5]In the four-qubit case, the ring of SLOCC invariant polynomials is generated by four polynomials denoted by H, L, M and D in [57]. One way of defining the quotient map is to consider $\Phi : \mathcal{H} \to \mathbb{C}^4$ defined by $\Phi(\hat{x}) = (H(\hat{x}), L(\hat{x}), N(\hat{x}), D(\hat{x}))$, see [37].

[6]The discriminant of the miniversal deformation of a singularity parametrizes all singular deformations of the singularity [3].

corresponds to k bosonic n-qudit states. Geometrically, the highest weight orbit is the so-called Veronese embedding of $\mathbb{P}^{n-1}$ [29]:

$$v_k : \mathbb{P}^{n-1} \to \mathbb{P}(\mathrm{Sym}^k(\mathbb{C}^n))$$
$$[\psi] \mapsto [\underbrace{\psi \circ \psi \circ \cdots \circ \psi}_{k \text{ times}}]. \tag{5.28}$$

The variety $v_k(\mathbb{P}^{n-1}) \subset \mathbb{P}(\mathrm{Sym}^k(\mathbb{C}^n))$ is geometrically the analogue of the variety of separable states for multiqudit systems given by the Segre embedding. It is not completely clear what entanglement physically means for bosonic systems. The ambiguity comes from the fact that symmetric states like $|W_3\rangle = \frac{1}{3}(|100\rangle + |010\rangle + |001\rangle)$ can be factorized under the symmetric tensor product $|W_3\rangle = |1\rangle \circ |0\rangle \circ |0\rangle$. However we can define entanglement in such symmetric systems by considering the space of symmetric states as a subset of the space of k n-dits states $\mathbb{C}^n \otimes \cdots \otimes \mathbb{C}^n$. In this case the kth-Veronese embedding of $\mathbb{P}^{n-1}$ corresponds to the intersection of the variety of separable states $\mathbb{P}^{n-1} \times \cdots \times \mathbb{P}^{n-1}$ with $\mathbb{P}(\mathrm{Sym}^k(\mathbb{C}^n))$ [16]. In the special case of $n = 2$, the variety $v_k(\mathbb{P}^1) \subset \mathbb{P}^k$ can also be identified with the variety of spin s-coherent states $(2s = k)$ when a spin s-state is given as a collection of $2s$ spin $\frac{1}{2}$-particles [7, 21]. For a comprehensive study about entanglement of symmetric states, see [6].

- Consider the simple complex Lie group SLOCC $= \mathrm{SL}_n(\mathbb{C})$ and its irreducible representation $\mathcal{H}_{fermions} = \bigwedge^k \mathbb{C}^n$ which is the the space of skew symmetric k tensors over $\mathbb{C}^n$. This Hilbert space represents the space of k skew-symmetric particles with n-modes, i.e. k fermions with n-single particle states. In this case the highest weight orbit is also a well-known algebraic variety, called the Grassmannian variety $G(k, n)$. The Grassmannian variety $G(k, n)$ is the set of k planes in $\mathbb{C}^n$ and it is defined as a subvariety of $\mathbb{P}(\bigwedge^k \mathbb{C}^n)$ by the Plücker embedding [29]:

$$G(k, n) \qquad\qquad \hookrightarrow \mathbb{P}(\textstyle\bigwedge^k \mathbb{C}^n)$$
$$\mathrm{Span}\{v_1, v_2, \ldots, v_k\} \mapsto [v_1 \wedge v_2 \wedge \cdots \wedge v_k]. \tag{5.29}$$

From the point of view of quantum physics the Grassmannian variety represents the set of fermions with Slater rank one and is naturally considered as the set of non-entangled states.

Another type of quantum system which can be described by means of representation theory is the case of particles in a fermionic Fock space with finite N-modes [74]. A fermionic Fock space with finite N-modes physically describes fermionic systems with N-single particle states, where the number of particles is not necessarily conserved by the admissible transformations. Let us recall the basic ingredient to describe such a Hilbert space. Let V be an $N = 2n$-dimensional complex vector space corresponding to one particle states. The associated fermionic Fock space is given by:

$$\mathcal{F} = \wedge^{\bullet} V = \mathbb{C} \oplus V \oplus \wedge^2 V \oplus \cdots \oplus \wedge^N V = \underbrace{\wedge^{even} V}_{\mathcal{F}_+} \oplus \underbrace{\wedge^{odd} V}_{\mathcal{F}_-}. \tag{5.30}$$

Similarly to the bosonic Fock space description of the Harmonic oscillator, one may describe this vector space as generated from the vacuum $|0\rangle$ (a generator of $\wedge^0 V$) by applying creation operators $\mathbf{p}_i$, $1 \leq i \leq N$. Thus a state $|\psi\rangle \in \mathcal{F}$ is given by

$$|\psi\rangle = \sum_{i_1,\dots,i_k} \psi_{i_1,\dots,i_k} \mathbf{p}_{i_1} \cdots \mathbf{p}_{i_k} |0\rangle \text{ with } \psi_{i_1,\dots,i_k} \text{ skew symmetric tensors.}$$

$$\tag{5.31}$$

The annihilation operators $\mathbf{n}_j$, $1 \leq j \leq N$ are defined such that $\mathbf{n}_j |0\rangle = 0$ and satisfy the Canonical Anticommutation Relations (CAR)

$$\{\mathbf{p}_i, \mathbf{n}_j\} = \mathbf{p}_i \mathbf{n}_j + \mathbf{n}_j \mathbf{p}_i = \delta_{ij}, \{\mathbf{p}_i, \mathbf{p}_j\} = 0, \{\mathbf{n}_i, \mathbf{n}_j\} = 0. \tag{5.32}$$

To see the connection with Lie group representation, let us consider $W = V \oplus V'$ where V and V' are isotropic subspaces, with basis $(e_j)_{1 \leq j \leq 2N}$, for the quadratic form $Q = \begin{pmatrix} 0 & I_N \\ I_N & 0 \end{pmatrix}$ and let us denote by $Cl(W, Q)$ the corresponding Clifford algebra [27]. Thus $\mathcal{F}$ is a $Cl(W, Q)$ module

$$w = x_i e_i + y_j e_{N+j} \mapsto \sqrt{2}(x_i \mathbf{p}_i + y_j \mathbf{n}_j) \in End(\mathcal{F}). \tag{5.33}$$

It follows that $\mathcal{F}_+$ and $\mathcal{F}_-$ are irreducible representations of the simple Lie group Spin$(2N)$, i.e. the spin group.[7] Those irreducible representations are known as spinor representations.

Example (The Box Picture) Let $V = \mathbb{C}^{2n} = \mathbb{C}^2 \otimes \mathbb{C}^n$, i.e. a single particle can be in two different modes ($\uparrow$ or $\downarrow$) and n different locations. We denote by $\mathbf{p}_1, \dots, \mathbf{p}_n, \mathbf{p}_{\bar{1}}, \dots, \mathbf{p}_{\bar{n}}$ the corresponding creation operators where $\mathbf{p}_i$ creates an $\uparrow$-particle in the i-th location and $\mathbf{p}_{\bar{i}}$ creates a $\downarrow$-particle in the i-th location. One can give a box picture representation of the embedding of n qubits in the Hilbert space $\mathcal{F} = \mathcal{F}_+ \oplus \mathcal{F}_-$. With the chirality decomposition $\mathcal{F} = \mathcal{F}_+ \oplus \mathcal{F}_-$ one gets two different ways of embedding n qubits, Figs. 5.6 and 5.7.

If we consider quantum information processing involving n bosonic qubits, n qubits or n fermions with $2n$ modes, all systems can be naturally embedded in the

---

[7]The spin group Spin$(2N)$ corresponds to the simply connected double cover of SO$(2N)$ [27].

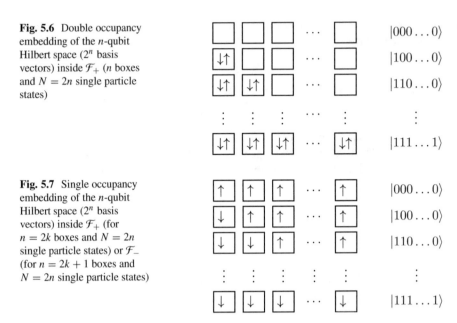

Fig. 5.6 Double occupancy embedding of the n-qubit Hilbert space (2^n basis vectors) inside $\mathcal{F}_+$ (n boxes and $N = 2n$ single particle states)

Fig. 5.7 Single occupancy embedding of the n-qubit Hilbert space (2^n basis vectors) inside $\mathcal{F}_+$ (for $n = 2k$ boxes and $N = 2n$ single particle states) or $\mathcal{F}_-$ (for $n = 2k + 1$ boxes and $N = 2n$ single particle states)

fermionic Fock space with $N = 2n$ modes and the restriction of the action of the Spin($2N$) = Spin($4n$) group to those sub-Hilbert-spaces boils down to their natural SLOCC group as shown in Table 5.2. In this sense the Spin group can be regarded as a natural generalization of the SLOCC group.

Let us denote by Δ_{4n} the irreducible representations $\mathcal{F}_\pm$, the algebraic variety $\mathbb{S}_{2n} \subset \mathbb{P}(\Delta_{4n})$ corresponding to the highest weight orbit of Spin($4n$) is called the spinor variety and generalizes the set of separable states. Table 5.2 indicates that the classification of spinors could be considered as the general framework to study the entanglement classification of pure quantum systems. The embedding of qubits into fermionic systems (with a fixed number of particles) was used in [20] to answer the question of SLOCC equivalence in the four-qubit case. In [51] we used the embedding within the fermionic Fock space to recover the polynomial invariants of the four-qubit case from the invariants of the spinor representation.

5.2.5 From Sequence of Simple Lie Algebras to the Classification of Tripartite Quantum Systems with Similar Classes of Entanglement

Let us go back to the three qubits classification and the $|W_3\rangle$ and $|GHZ_3\rangle$ states. After the paper of Dür, Vidal and Cirac [25] other papers were published in the quantum information literature describing other quantum systems featuring only two types of genuine entangled states, similar to the $|W_3\rangle$- and $|GHZ_3\rangle$-states.

Table 5.2 Embedding of n-bosonic qubit, n-qubit, n fermions with $2n$ single particle states into fermionic Fock space with $2N = 4n$ modes

Lie algebra	$\mathfrak{sl}_2$		$\mathfrak{sl}_2 + \cdots + \mathfrak{sl}_2$		$\mathfrak{sl}_{2n}$		$\mathfrak{so}_{4n}$
Lie group	$\mathrm{SL}_2(\mathbb{C})$	$\subset$	$\mathrm{SL}_2(\mathbb{C}) \times \cdots \times \mathrm{SL}_2(\mathbb{C})$	$\subset$	$\mathrm{SL}_{2n}(\mathbb{C})$	$\subset$	$\mathrm{Spin}(4n)$
Representation	$\mathbb{P}(\mathrm{Sym}^n(\mathbb{C}^2))$	$\hookrightarrow$	$\mathbb{P}(\mathbb{C}^2 \otimes \cdots \otimes \mathbb{C}^2)$	$\hookrightarrow$	$\mathbb{P}(\bigwedge^n \mathbb{C}^{2n})$	$\hookrightarrow$	$\mathbb{P}(\Delta_{4n})$
Highest weight orbit	$v_n(\mathbb{P}^1)$	$\subset$	$\mathbb{P}^1 \times \cdots \times \mathbb{P}^1$	$\subset$	$G(n, 2n)$	$\subset$	$\mathbb{S}_{2n}$

In [33] we showed how all those similar classifications correspond to a sequence of varieties studied from representation theory and algebraic geometry in connection with the Freudenthal magic square [46]. Consider a Lie group G acting by its adjoint action on its Lie algebra $\mathfrak{g}$. The adjoint variety $X_G \subset \mathbb{P}(\mathfrak{g})$ is the highest weight orbit for the adjoint action. Take any point $x \in X_G$ and let us consider the set of all lines of X_G passing through x (these lines are tangent to X_G). This set of lines is a smooth homogeneous variety $Y \subset \mathbb{P}(T_x X_G)$, called the subadjoint variety of X_G. Consider the sequence of Lie algebras

$$\mathfrak{g}_2 \subset \mathfrak{so}_8 \subset \mathfrak{f}_4 \subset \mathfrak{e}_6 \subset \mathfrak{e}_7. \tag{5.34}$$

This sequence gives rise to a series of subadjoint varieties called the *subexceptional series*. In [46] this sequence is obtained as the third row of the geometric version of the Freudenthal's magic square.

To see how the subexceptional series is connected to the different classifications of [11, 16, 24, 50, 53, 74] let us ask the following question: What do the Hilbert spaces $\mathcal{H}$ and the corresponding SLOCC groups G look like when the only genuine entanglement types are $|W\rangle$ and $|GHZ\rangle$?

If we assume that G is a Lie group and $\mathcal{H}$ an irreducible representation such that the only two types of genuine entangled states are $|W\rangle$ and $|GHZ\rangle$ then one knows from Sect. 5.2.2 that the secant variety of the variety of separable states should fill the ambient space and be of the expected dimension. Because the secant variety is an orbit, this orbit is dense by our assumption and, therefore, the ring of SLOCC invariant polynomials should be generated by at most one element. But one also knows, under our assumption and by Zak's theorem, that in this case the tangential variety, i.e. the $|W\rangle$-orbit, is a codimension-one orbit in the ambient space. Thus the ring of G-invariant polynomials for the representation $\mathcal{H}$ should be generated by a unique polynomial. The classification of such representations was given in the 70s by Kac, Popov and Vinberg [42]. From this classification one just needs to keep the representation where the dimension of the secant variety of the highest weight orbit is of the expected dimension. This leads naturally to the sequence of subexceptional varieties as given in Table 5.3.

Remark 5.4 The relation between the Freudenthal magic square and the tripartite entanglement was already pointed out in [11, 79]. Other subadjoint varieties for the Lie algebra $\mathfrak{so}_{2n}$, $n \neq 4$, not included in the subexceptional series also share the same orbit structure. The physical interpretation of those systems is clear for $n = 3, 5, 6$ [33, 79], but rather obscure in the general case $n \geq 7$.

Remark 5.5 This sequence of systems can also be considered from the dual picture by looking for generalization of the Cayley hyperdeterminant (the dual equation of $X = \mathbb{P}^1 \times \mathbb{P}^1 \times \mathbb{P}^1$). In [46] it was also shown that all dual equations for the subexceptional series can be uniformly described. The tripartite entanglement of seven qubits [24], under constrains given by the Fano plane, also started with a generalization of Cayley's quartic hyperdeterminant in relation with black-hole-entropy formulas in the context of the black-hole/qubit correspondence [13].

Table 5.3 The sequence of subexceptional varieties and the corresponding tripartite systems

$\mathcal{H}$	SLOCC	QIT interpretation	$X_{Sep} \subset \mathbb{P}(\mathcal{H})$	$\mathfrak{g}$
$Sym^3(\mathbb{C}^2)$	$SL_2(\mathbb{C})$	Three bosonic qubit [16, 79] (2007)	$v_3(\mathbb{P}^1) \subset \mathbb{P}^3$	$\mathfrak{g}_2$
$\mathbb{C}^2 \otimes \mathbb{C}^2 \otimes \mathbb{C}^2$	$SL_2(\mathbb{C}) \times SL_2(\mathbb{C}) \times SL_2(\mathbb{C})$	Three qubit [25] (2001)	$\mathbb{P}^1 \times \mathbb{P}^1 \times \mathbb{P}^1 \subset \mathbb{P}^7$	$\mathfrak{so}_8$
$\bigwedge^{(3)} \mathbb{C}^6$	$Sp_6(\mathbb{C})$	Three fermions with with six single particles state with a symplectic condition	$LG(3, 6) \subset \mathbb{P}^{13}$	$\mathfrak{f}_4$
$\bigwedge^3 \mathbb{C}^6$	$SL_6(\mathbb{C})$	Three fermions with with six single particles state [53] (2008)	$G(3, 6) \subset \mathbb{P}^{19}$	$\mathfrak{e}_6$
Δ_{12}	$Spin(12)$	Particles in Fermionic Fock space [74] (2014)	$\mathbb{S}_6 \subset \mathbb{P}^{31}$	$\mathfrak{e}_7$
V_{56}	E_7	Three partite entanglement of seven qubit [24, 50] (2007)	$E_7/P_1 \subset \mathbb{P}^{55}$	$\mathfrak{e}_8$
		Freudenthal subexceptionnal series		

5.3   The Geometry of Contextuality

In this second part of the paper I discuss the finite geometry behind operator-based proofs of contextuality. Starting from the geometric description of the N-qubit Pauli group, I recall how the concept of Veldkamp geometry associated to a point line configuration recently leaded us to recognize weight diagrams of simple Lie algebras in some specific arrangement of hyperplanes of the three-qubit Pauli group.

5.3.1   Observable-Based Proofs of Contextuality

As explained in the introduction, operator-based proofs of the Kochen-Specker (KS) Theorem correspond to configurations of mutli-Pauli observables such that the operators on the same context (line) are mutually commuting and such that the product of the operators gives $\pm I$, with an odd number of negative contexts.

The Mermin-Peres square presented in the introduction is the first operator/ observable-based proof of the KS Theorem. In [59], Mermin also proposed another proof involving three qubit Pauli operators and known as the Mermin pentagram (Fig. 5.8).

The Mermin-Peres square and the Mermin pentagram are the smallest configurations, in terms of number of contexts and number of operators, providing observable based proofs of contextuality [34]. Other proofs of the KS Theorem based on observable configurations have been proposed by Waegel and Aravind [81, 82] or Planat and Saniga [66, 70]. In terms of quantum processing, the "magic" configurations have been investigated under the scope of non-local games. For each magic configuration one can define a game where cooperative players can win with certainty using a quantum strategy. Let us look at the magic square of Fig. 5.2 and consider the following game involving two players Alice and Bob, and a referee

Fig. 5.8 The Mermin pentagram: a configuration of 10 three-qubit operators proving KS Theorem. Operators on a line are mutually commuting and the doubled line corresponds to the context where the product gives $-I_8$

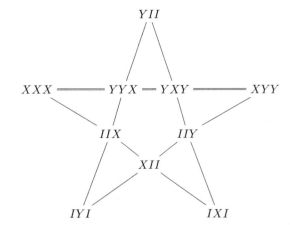

Charlie. As usual, Alice and Bob may define a strategy in advance but cannot communicate once the game starts:

1. Charlie picks a number $r \in \{1, 2, 3\}$ for a row and $c \in \{1, 2, 3\}$ for a column and sends r to Alice and c to Bob.
2. Both Alice and Bob send back to the referee a triplet of ± 1 such that the number of -1 is odd for Alice and even for Bob.
3. Alice and Bob win the game if the number in position c of Alice triplet matches with the number in position r for Bob's triplet (and of course the triplets of Alice and Bob satisfy the parity condition of the previous step).

Such type of game is called a binary constrain game [23]. If Alice and Bob share a specific four-qubit entangled state (a product of two $|EPR\rangle$-like states) they can win that game with certainty, while it is easy to prove that there is no such classical strategy. In [2], Arkhipov gave a graph-theoretic characterization of magic configurations in terms of planarity of the dual configuration.

A natural question to ask is to find all possible different realizations of a given magic configuration. For instance one can ask how many two-qubit KS proof similar to the Mermin-Peres square can be built, or how many Mermin pentagrams can we obtain with three-qubit Pauli operators? As we will explain now this can be answered by looking at the geometry of the space of N-qubit Pauli operators.

Remark 5.6 Originally, the first proof of KS was not given in terms of configurations of multiqubit Pauli-operators, but by considering projection operators on some specific basis of the three-dimensional Hilbert space. Kochen and Specker found a set of 117 operators and proved the impossibility to assign a deterministic value ± 1 to each of them by using a coloring argument on the corresponding basis vectors. Several simplification of this original proof were proposed in the literature. For instance, one can reduce to 18 the number of vectors needed to express the KS Theorem in terms of projectors [19].

5.3.2 The Symplectic Polar Space of Rank N and the N-Qubit Pauli Group

To understand where these magic configurations live, we now start to describe geometrically the generalized N-qubit Pauli group, i.e. the group of Pauli operators acting on N-qubit systems. The following construction is due to M. Saniga and M. Planat [30, 69, 77] and has been employed in the past 10 years to provide a finite geometric insight starting from the commutation relations of Pauli observables up to the black-hole-entropy formulas [13, 54, 55].

Let us consider the subgroup P_N of $GL(2^N, \mathbb{C})$ generated by the tensor products of Pauli matrices,

$$A_1 \otimes A_2 \otimes \cdots \otimes A_N \equiv A_1 A_2 \ldots A_N, \tag{5.35}$$

with $A_i \in \{\pm I, \pm iI, \pm X, \pm iX, \pm Y, \pm iY, \pm Z, \pm iZ\}$. The center of P_N is $C(P_N) = \{\pm I, \pm iI\}$ and $V_N = P_N/C(P_N)$ is an abelian group.

To any class $\overline{O} \in V_N$, there corresponds a unique element in $\mathbb{F}_2^{2N}$. More precisely, for any $O \in P_N$ we have $O = sZ^{\mu_1}X^{\nu_1} \otimes \cdots \otimes Z^{\mu_N}X^{\nu_N}$ with $s \in \{\pm 1, \pm i\}$ and $(\mu_1, \nu_1, \ldots, \mu_N, \nu_N) \in \mathbb{F}_2^{2N}$. Thus V_N is a $2N$ dimensional vector space over $\mathbb{F}_2$ and we can associate to any non-trivial observable $O \in P_N \setminus I^N$ a unique point in the projective space $\mathbb{P}_2^{2N-1} = \mathbb{P}(\mathbb{F}_2^{2N})$.

$$\begin{aligned} \pi : P_N \setminus I_N &\rightarrow \mathbb{P}_2^{2N-1} \\ O = sZ^{\mu_1}X^{\nu_1} \otimes \cdots \otimes Z^{\mu_N}X^{\nu_N} &\mapsto [\mu_1 : \nu_1 : \cdots : \mu_N : \nu_N]. \end{aligned} \tag{5.36}$$

Because V_N is a vector space over $\mathbb{F}_2$, the lines of $\mathbb{P}_2^{2N-1}$ are made of triplet of points (α, β, γ) such that $\gamma = \alpha + \beta$. The corresponding (class) of observables $\overline{O_\alpha}$, $\overline{O_\beta}$ and $\overline{O_\gamma}$ satisfy $\overline{O_\alpha}.\overline{O_\beta} = \overline{O_\gamma}$ (. denotes the ordinary product of operators).

Example For single qubit we have $\pi(X) = [0:1]$, $\pi(Y) = [1:1]$ and $\pi(Z) = [1:0]$. The projective space $\mathbb{P}_2^1$ is the projective line (X, Y, Z) (the projection π will be omitted).

However, the correspondence between non-trivial operators of P_N and points in $\mathbb{P}_2^{2N-1}$ does not say anything about the commutation relations between the operators. To see geometrically these commutation relations, one needs to introduce an extra structure. Let $O, O' \in P_N$ such that $O = sZ^{\mu_1}X^{\nu_1} \otimes \cdots \otimes Z^{\mu_N}X^{\nu_N}$ and $O' = s'Z^{\mu'_1}X^{\nu'_1} \otimes \cdots \otimes Z^{\mu'_N}X^{\nu'_N}$ with $s, s' \in \{\pm 1, \pm i\}$ and $\mu_i, \nu_i, \mu'_i, \nu'_i \in \mathbb{F}_2$.

Then, we have

$$O.O' = (ss'(-1)^{\sum_{j=1}^N \mu'_j \nu_j}, \mu_1 + \nu'_1, \ldots, \mu_N + \nu'_N), \tag{5.37}$$

and the two elements O and O' of P_N commute, if and only, if

$$\sum_{j=1}^N (\mu_j \nu'_j + \mu'_j \nu_j) = 0. \tag{5.38}$$

Let us add to V_N the symplectic form

$$\langle \overline{O}, \overline{O'} \rangle = \sum_{j=1}^N (\mu_j \nu'_j + \mu'_j \nu_j), \tag{5.39}$$

and let us denote by $\mathcal{W}(2N-1, 2)$ the symplectic polar space of rank N i.e. the set of totally isotropic subspaces of $(\mathbb{P}_2^{2N-1}, \langle, \rangle)$. The symplectic polar space $\mathcal{W}(2N-1, 2)$ encodes the commutation relations of $P_N \setminus I_N$. The points of $\mathcal{W}(2N-1, 2)$ correspond to non trivial operators of P_N and the subspaces of $\mathcal{W}(2N-1, 2)$ correspond to $\mathbb{P}(S/C(P_N))$, where S is a set of mutually commuting elements of P_N.

5.3.3   Geometry of Hyperplanes: Veldkamp Space of a Point-Line Geometry

The points and lines of $\mathcal{W}(2N-1,2)$ define an incidence structure, i.e. a point-line geometry $G = (\mathcal{P}, \mathcal{L}, \mathcal{I})$ where $\mathcal{P}$ are the points of $\mathcal{W}(2N-1,2)$, $\mathcal{L}$ are the lines and $\mathcal{I} \subset \mathcal{P} \times \mathcal{L}$ corresponds to the incidence relation. I now introduce some geometric notions for point-line incidence structures.

Definition 5.4 Let $G = (\mathcal{P}, \mathcal{L}, \mathcal{I})$ be a point-line incidence structure. A hyperplane H of G is a subset of $\mathcal{P}$ such that a line of $\mathcal{L}$ is either contained in H, or has a unique intersection with H.

Example Let us consider a 3×3 grid with 3 points per line, also known as $GQ(2, 1)$. This geometry has 15 hyperplanes splitting in two different types: the perp sets (the unions of two "perpendicular" lines) and the ovoids (hyperplanes that contain no lines), see Fig. 5.9.

The notion of geometric hyperplanes leads to the notion of Veldkamp space as introduced in [71].

Definition 5.5 Let $G = (\mathcal{P}, \mathcal{L}, \mathcal{I})$ be a point-line geometry. The Veldkamp space of G, denoted by $\mathcal{V}(G)$, if it exists, is a point-line geometry such that

- the points of $\mathcal{V}(G)$ are geometric hyperplanes of G,
- given two points H_1 and H_2 of $\mathcal{V}(G)$, the Veldkamp line defined by H_1 and H_2 is the set of hyperplanes of G such that $H_1 \cap H = H_2 \cap H$ or $H = H_i, i = 1, 2$.

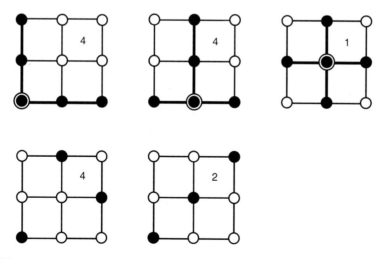

Fig. 5.9 Pictural representation of the 15 hyperplanes of the grid $GQ(2, 1)$. Nine hyperplanes are of type perp and six of them are of type ovoid

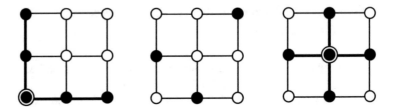

Fig. 5.10 An example of Veldkamp line of the grid, i.e. a line of $\mathcal{V}(GQ(2,1))$. The three hyperplanes share two by two the same intersection (and no other hyperplane of $GQ(2,1)$ does). Taking two of those three hyperplanes, the thrid one is obtained by considering the complement of the symmetric difference (see Eq. (5.40) below)

Figure 5.10 furnishes an example of a Veldkamp line in $\mathcal{V}(GQ(2,1))$. It is not too difficult to show that the hyperplanes of $GQ(2,1)$ accommodate the 15 points and 35 lines of $\mathbb{P}_2^3$, i.e. $\mathcal{V}(GQ(2,1)) = \mathbb{P}_2^3$.

Remark 5.7 The notion of Veldkamp space of finite point-line incidence structures has been employed to study orbits in $\mathbb{P}_2^{2^N-1}$ under the action of $SL_2(\mathbb{F}_2) \times \cdots \times SL_2(\mathbb{F}_2)$. For $N = 4$, it was possible to obtain a computer free proof of the classification of the $2 \times 2 \times 2 \times 2$ tensors over $\mathbb{F}_2$ by classifying the hyperplanes of a specific configuration. More precisely it was shown in [72] that the Veldkamp space of the finite Segre varieties of type $S_N = \underbrace{\mathbb{P}_2^1 \times \cdots \times \mathbb{P}_2^1}_{N \text{ times}}$ is the projective space $\mathbb{P}_2^{2^N-1}$ and that the different types of hyperplanes of S_N are in bijection with the $SL_2(\mathbb{F}_2) \times \cdots \times SL_2(\mathbb{F}_2)$-orbits of $\mathbb{P}_2^{2^N-1}$.

5.3.4 The Finite Geometry of the Two-Qubit and Three-Qubit Pauli Groups and the Hyperplanes of $\mathcal{W}(2N-1, 2)$

We now describe in detail $\mathcal{W}(3, 2)$ and $\mathcal{W}(5, 2)$, the symplectic polar spaces encoding the commutation relations of the two and three-qubit Pauli groups and their Veldkamp spaces.

The symplectic polar space $\mathcal{W}(3, 2)$ consists of all 15 points of $\mathbb{P}_2^3$ but only the 15 isotropic lines are kept. This gives a point-line configuration description of $\mathcal{W}(3, 2)$, Fig. 5.11, known as the doily [30].

The doily is also known as the generalized quadrangle[8] $GQ(2, 2)$. In the following I will keep denoting by $\mathcal{W}(3, 2)$ both the symplectic polar space and the

---

[8] A point-line incidence structure is called a generalized quadrangle of type (s, t), and denoted by $GQ(s, t)$ iff it is an incidence structure such that every point is on $t+1$ lines and every line contains $s + 1$ points such that if $p \notin L$, $\exists! q \in L$ such that p and q are collinear.

Fig. 5.11 The labeling of the doily, i.e. the symplectic polar space $\mathcal{W}(3, 2)$, by Pauli operators. The doily is a 15_3-configuration (15 points, 15 lines, 3 points per line and 3 lines through each point) which is a generalized quadrangle (i.e. is triangle free). It is the unique 15_3-configuration that is triangle free among $245,342$ ones. The doily encodes the commutation relations of the two-qubit Pauli group

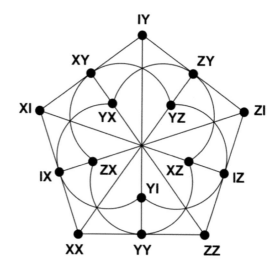

associated point-line geometry $GQ(2, 2)$. Looking at the doily (Fig. 5.11) one can identify the Mermin-Peres squares built with two-qubit Pauli operators as geometric hyperplanes of $\mathcal{W}(3, 2)$.

In fact, three different types of hyperplanes can be found in the doily as shown in Fig. 5.12:

- The hyperplanes made of 9 points (red) correspond to grids $GQ(2, 1)$ and it is easy to check that grids on the two-qubit Pauli group are always contextual configurations [34], i.e. Mermin-Peres squares. Rotating by $\dfrac{2\pi}{5}$ one gets 10 Mermin-Peres grids in the doily.

- The second type of hyperplanes (yellow ones) are called perp-sets (all lines of the hyperplane meet in one point) and one sees from Fig. 5.12 that there are 15 of such.

- Finally the last type of hyperplanes of the doily (blue) are line-free and such type of hyperplanes are called ovoids. The doily contains six ovoids.

The geometry of $\mathcal{V}(GQ(2, 2))$, the Veldkamp space of the doily, is described in full details in [71]. Figure 5.13 illustrates the different types of Veldkamp lines that can be obtained from the hyperplanes of the doily.

In particular, $\mathcal{V}(\mathcal{W}(3, 2))$ comprises 31 points splitting in three orbits and 155 lines splitting in five different types. One can show that $\mathcal{V}(GQ(2, 2)) \simeq \mathbb{P}_2^4$.

The symplectic polar space $\mathcal{W}(5, 2)$ contains 63 points, 315 lines, 135 Fano planes. One can build 12,096 distinguished Mermin pentagrams from those 63 points [55, 67].

In the case of the three-qubit Pauli group there is no generalized polygon which accommodates the full geometry $\mathcal{W}(5, 2)$. However, there exists an embedding in $\mathcal{W}(5, 2)$ of the split-Cayley hexagon of order two, which is a generalized hexagon of 63 points and 63 lines such that each line contains 3 points and each point belongs to

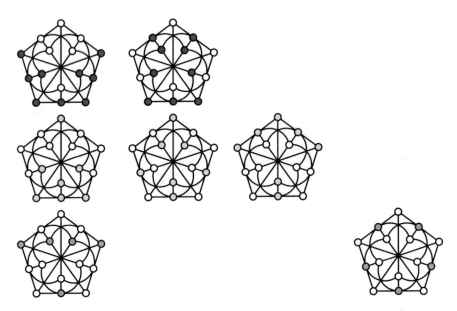

Fig. 5.12 The three different types of hyperplanes of the doily [71]. In red hyperplanes corresponding to grids, $GQ(2, 1)$, in yellow hyperplanes corresponding to perp-sets and in blue hyperplanes of type ovoids

3 lines. This split-Cayley hexagon accommodates the 63 three-qubit operators of the three-qubit Pauli group such that the lines of the configuration are totally isotropic lines (Fig. 5.14).

The general structure of $\mathcal{V}(\mathcal{W}(2N - 1, 2))$ has been studied in details in [80] where the description of the geometric hyperplanes of $\mathcal{W}(2N - 1, 2)$ is explicitly given. First, let us mention that for $\mathcal{G} = \mathcal{W}(2N - 1, 2)$ the Veldkamp line defined by two hyperplanes H_1 and H_2 is a 3-point line (H_1, H_2, H_3) where H_3 is given by the complement of the symmetric difference,

$$H_3 = H_1 \boxplus H_2 = \overline{H_1 \Delta H_2}. \tag{5.40}$$

To reproduce the description of $\mathcal{V}(\mathcal{W}(2N - 1, 2))$ of [80], let us introduce the following quadratic form over V_N:

$$Q_0(x) = \sum_{i=1}^{N} a_i b_i \text{ where } x = (a_1, b_1, \ldots, a_N, b_N). \tag{5.41}$$

An observable O is said to be symmetric if it contains an even number of Y's or skew-symmetric if it contains an odd number of Y's. In terms of the quadratic form Q_0, this leads to the conditions $Q_0(\overline{O}) = 0$ or $Q_0(\overline{O}) = 1$.

Fig. 5.13 The five types of Veldkamp lines of the doily [71]. For each line, the points collored in black correspond to the core of the Velkamp line. Note that two types of lines (the second and third) have the same composition (perp-perp-perp) and are distinguished by the core set, which is either composed of three noncolinear points or three points on a line. One can check that given any two hyperplanes on a line, the third one is the complement of the symmetric difference of the two, see Eq. (5.40)

There are three types of geometric hyperplanes in $\mathcal{W}(2N-1, 2)$:

$$\text{Type 1: } C_q = \{p \in \mathcal{W}(2N-1, 2), \langle p, q \rangle = 0\}. \tag{5.42}$$

This set corresponds to the "perp-set" defined by q, i.e. in terms of operators, it is the set of elements commuting with $\overline{O}_q$.

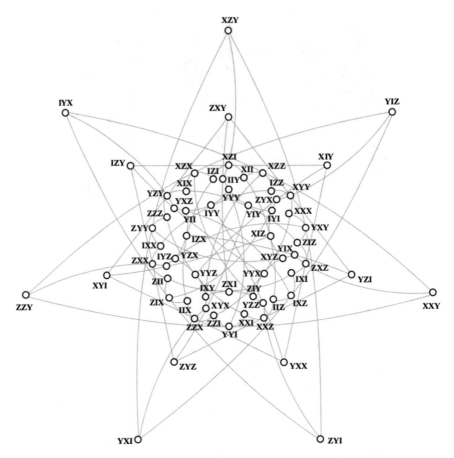

Fig. 5.14 A three-qubit Pauli group embedding into the split Cayley hexagon [54] in $\mathcal{W}(5, 2)$. The split Cayley hexagon is a generalized polygon, it is a 63_3 configuration that contains no ordinary pentagon

To define Type 2 and Type 3, let us introduce a family of quadratic forms on V_N parametrized by the elements of V_N: $Q_q(p) = Q_0(p) + \langle q, p \rangle$. Depending on the nature of $\overline{O}_q$ (symmetric or skew-symmetric) the quadratic form will be hyperbolic or elliptic.

Type 2: for $\overline{O}_q$ symmetric $H_q = \{p \in \mathcal{W}(2N - 1, 2), Q_q(p) = 0\} \simeq Q^+(2N - 1, 2)$,

$$(5.43)$$

and

Type 3: for $\overline{O}_q$ skew-symmetric $H_q = \{p \in \mathcal{W}(2N - 1, 2), Q_q(p) = 0\} \simeq Q^-(2N - 1, 2)$,

$$(5.44)$$

where $Q^+(2N - 1, 2)$ denotes a hyperbolic quadric[9] of $\mathcal{W}(2N - 1, 2)$, and $Q^-(2N - 1, 2)$ denotes an elliptic quadric[10] of $\mathcal{W}(2N - 1, 2)$.

The set H_q represents the set of observables either symmetric and commuting with $\overline{O}_q$ or skew-symmetric and anticommuting with $\overline{O}_q$.

Moreover, the following equalities hold

$$C_p \boxplus C_q = C_{p+q}, \; H_p \boxplus H_q = C_{p+q} \text{ and } C_p \boxplus H_q = H_{p+q}. \tag{5.45}$$

This leads to five different types of Veldkamp lines in $\mathcal{W}(2N - 1, 2)$ depending on the nature (symmetric or not) of the points p and q (we recover the five different types of Veldkamp lines illustrated in Fig. 5.13).

5.3.5 From Commutation Relations of the Three-Qubit Pauli Group to the Weight Diagrams of Simple Lie Algebras

It was first pointed out in [52] that the Mermin pentagrams showing up in the three-qubit Pauli group can all be obtained from a "double six" configuration of such pentagrams living in a Veldkamp line of type perp-hyperbolic-elliptic. More precisely, taking the transitive action of the symplectic group Sp(6, 2) on $\mathcal{W}(5, 2)$, one can recover all Mermin pentagrams from the 12 pentagrams living in a specific subspace of the Veldkamp line $(H_{III}, H_{YYY}, C_{YYY})$. According to the previous subsection one has the following description of the three hyperplanes H_{III}, H_{YYY} and C_{YYY} in terms of Pauli operators,

- C_{YYY} is the perp-set defined by the operator YYY, i.e. the points in C_{YYY} correspond to operators commuting with YYY.
- H_{III} is a hyperbolic quadric, i.e. is defined by $Q_0(x) = 0$. In terms of operators it corresponds to the set of symmetric operators (i.e. containing an even number of Y).
- H_{YYY} is an elliptic quadric, i.e. is defined by $Q_{YYY}(x) = 0$. In terms of operators it corresponds to the set of symmetric operators commuting with YYY or skew-symmetric ones anti-commuting with YYY.

The core set of the Veldkamp line is the set of elements commuting with YYY (they belong to C_{YYY}) and symmetric (they belong to H_{III}). An explicit list of those elements is given by:

$$\begin{aligned} YYI \;\; YIY \;\; IYY \;\; ZZI \;\; ZIZ \;\; IZZ \;\; XXI \;\; XIX \\ IXX \;\; ZXI \;\; ZIX \;\; IZX \;\; XZI \;\; XIZ \;\; IXZ. \end{aligned} \tag{5.46}$$

---

[9]Up to a transformation of coordinates, this is a set of points $x \in \mathbb{P}_2^{2N-1}$ satisfying the standard equation $x_1x_2 + x_3x_4 + \cdots + x_{2N-1}x_{2N} = 0$.

[10]Up to a transformation of coordinates this is defined as points $x \in \mathbb{P}_2^{2N-1}$ such that $f(x_1, x_1) + x_2x_3 + \ldots x_{2N-1}x_{2N} = 0$.

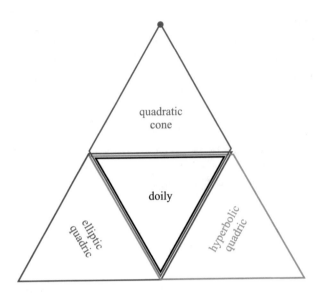

Fig. 5.15 Schematic representation of the Veldkamp line $(H_{III}, H_{YYY}, C_{YYY})$. The core set of this three-qubit Veldkamp line is made of 15 operators. The commutation relations among those 15 operators define the incidence structure of a doily

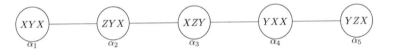

Fig. 5.16 Labelling of the Dynkin diagram of type A_5 by three-qubit Pauli operators

This set of observables forms a doily in $\mathcal{W}(5, 2)$ (see Fig. 5.15) that encapsulates the weight diagram of the second fundamental representation of A_5. To see this connection with simple Lie algebras, let us associate to the roots $\alpha_1, \ldots, \alpha_5$ of A_5 five skew-symmetric observables as given in Fig. 5.16. The action of the roots by translation on the weight vectors [27] corresponds to multiplication in terms of operators. Now taking ZIZ as the highest weight vector, then Fig. 5.17 reproduces the weight diagram of the 15-dimensional irreducible representation of A_5 built in terms of the three-qubit operators corresponding to the doily of Fig. 5.15.

This core set also encodes the Pfaffian of 6×6 skew-symmetric matrices which is the invariant of the 15-dimensional irreducible representation of A_5. To see this, consider the observable $\Omega = \sum_{1 \le i < j \le 6} a_{ij} O_{ij}$, where O_{ij} is a three-qubit observable located at (ij) (Fig. 5.18). Then the polynomial $Tr(\Omega^3)$ is proportional to the Pfaffian, $Pf(A)$, where $A = (a_{ij})_{1 \le i < j \le 6}$ is a skew symmetric matrix.

Remark 5.8 A different choice of representatives of the root system of A_5 will generate a different weight diagram with the operators composing the doily, i.e. the choice of the representatives of the roots determines the highest weight vector. In [56] we provided a labeling of the operators of the Veldkamp line in terms of a

Fig. 5.17 The weight diagram of the 15-dimensional representation of A_5 in terms of three-qubit operators. The action by the roots $\alpha_1, \ldots, \alpha_5$ of the Dynkin diagram is obtained by multiplying the weight by the three-qubit operator corresponding to the root (Fig. 5.16)

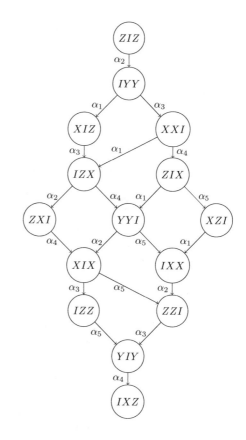

Fig. 5.18 Labeling of the doily by doublets. Two doublets are colinear if they have no element in common (the third doublet on the line being the complement of the two doublets)

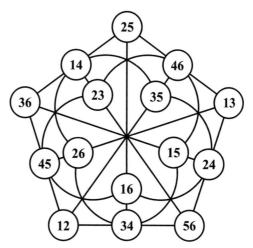

Table 5.4 Correspondence between hyperplanes, representations and invariants in the Veldkamp line $(C_{YYY}, H_{YYY}, H_{III})$

Geometry	Irreducible representation	Invariant
Quadratic cone	$1 \oplus 15 \oplus 15$ rep of A_5	Pfaffian (for the 15 of A_5)
Elliptic quadric	27 irrep of E_6	Cartan's cubic invariant
Hyperbolic quadric	35 irrep of A_6	7-order invariant

Clifford algebra. This has the double advantage to avoid a specific labeling but also establish a connection of the full Veldkamp line with the Spin(14) representation.

Similarly, all hyperplanes of the Veldkamp line $(C_{YYY}, H_{YYY}, H_{III})$ can be analyzed this way, revealing connection with the 27-dimensional irreducible representation of E_6 (elliptic quadric) or the 35-dimensional irreducible representation of A_6 (hyperbolic quadric) as well as their corresponding invariants (Table 5.4). Subparts (triangles in Fig. 5.15) can be combined to get other irreducible representations like the 32-dimensional irreducible representation of SO(12) which is made of the operators of the elliptic and hyperbolic quadrics which are not in the doily [56].

Remark 5.9 The hyperbolic quadric, i.e. the green part of the Veldkamp line Fig. 5.15, which corresponds to the weight diagram of the 35-dimensional irreducible representation of A_6, can be further decomposed as $35 = 15 \oplus 20$ for the action of A_5. In this decomposition the 15 of A_5 corresponds to the doily as detailed at the beginning of the section while the other symmetric operators, which accommodate the diagram of the 20-dimensional irreducible representation of A_5, generate the double-six of Mermin pentagrams [52].

Remark 5.10 In [56] other finite geometric structures, like extended quadrangle, are revealed in connection with sub-parts of this "magic" Veldkamp line.

5.4 Conclusion

The two geometric constructions presented in this paper have been known in the mathematics community for quite a long time. The concept of auxiliary varieties (secant, duals) has been known since the nineteenth century, while the notion of Veldkamp space of a point-line geometry goes back to the 80s of the twentieth century. These geometric constructions have been shown to be useful in quantum information in the past 15 years, first to describe quantum paradoxes such as entanglement and contextuality. These geometric approaches could be employed in the future to get insight into some quantum information protocols [39]. The fact that representation theory of simple Lie algebras acts as symmetry behind the scene could also lead to interesting findings of how to connect geometrically entanglement and contextuality.

Acknowledgements The first part of this review paper was presented at the international workshop "Quantum Physics and Geometry" organized at Levico Terme in July 2017. I would like to thank the organizers for inviting me to present an overview on my research and to contribute to this UMI Lecture Notes. The research presented in this review has been done collectively; I would like to warmly thank my co-authors Jean-Gabriel Luque, Jean-Yves Thibon, Michel Planat, Metod Saniga, Péter Lévay and Hamza Jaffali for our rich collaboration over the past 6 years. This work was partially supported by the French "Investissements d'Avenir" program, project ISITE-BFC (contract ANR-15-IDEX-03).

References

1. E. Amselem, M. Rådmark, M. Bourennane, A. Cabello, State-independent quantum contextuality with single photons. Phys. Rev. Lett. **103**(16), 160405 (2009)
2. A. Arkhipov, Extending and characterizing quantum magic games (2012). Preprint. arXiv:1209.3819
3. V.I. Arnold, Critical points of smooth functions, in *Proceedings of ICM-74*, vol. 1 (1974), pp. 19–40
4. V.I. Arnold, *Singularity Theory*, vol. 53 (Cambridge University Press, Cambridge, 1981)
5. A. Aspect, J. Dalibard, G. Roger, Experimental test of Bell's inequalities using time-varying analyzers. Phys. Rev. Lett. **49**(25), 1804 (1982)
6. M. Aulbach, Classification of entanglement in symmetric states. Int. J. Quantum Inf. **10**(07), 1230004 (2012)
7. M. Aulbach, D. Markham, M. Murao, Geometric entanglement of symmetric states and the majorana representation, in *TQC* (2010), pp. 141–158
8. H. Bartosik, J. Klepp, C. Schmitzer, S. Sponar, A. Cabello, H. Rauch, Y. Hasegawa, Experimental test of quantum contextuality in neutron interferometry. Phys. Rev. Lett. **103**(4), 040403 (2009)
9. J.S. Bell, On the problem of hidden variables in quantum mechanics. Rev. Mod. Phys. **38**(3), 447 (1966)
10. C.H. Bennett, S. Popescu, D. Rohrlich, J.A. Smolin, A.V. Thapliyal, Exact and asymptotic measures of multipartite pure-state entanglement. Phys. Rev. A 63(1), 012307 (2000)
11. L. Borsten, D. Dahanayake, M.J. Duff, W. Rubens, H. Ebrahim, Freudenthal triple classification of three-qubit entanglement. Phys. Rev. A 80(3), 032326 (2009)
12. L. Borsten, D. Dahanayake, M.J. Duff, A. Marrani, W. Rubens, Four-qubit entanglement classification from string theory. Phys. Rev. Lett. **105**(10), 100507 (2010)
13. L. Borsten, M.J. Duff, P. Lévay, The black-hole/qubit correspondence: an up-to-date review. Classical Quantum Gravity **29**(22), 224008 (2012)
14. E. Briand, J.G. Luque, J.Y. Thibon, A complete set of covariants of the four qubit system. J. Phys. A Math. Gen. **36**(38), 9915 (2003)
15. E. Briand, J.G. Luque, J.Y. Thibon, F. Verstraete, The moduli space of three-qutrit states. J. Math. Phys. **45**(12), 4855–4867 (2004)
16. D.C. Brody, A.C.T. Gustavsson, L.P. Hughston, Entanglement of three-qubit geometry. J. Phys. Conf. Ser. **67**, 012044 (2007)
17. J.L. Brylinski, Algebraic measures of entanglement, in *Mathematics of Quantum Computation* (Chapman and Hall/CRC, Boca Raton, 2002), pp. 3–23
18. A. Cabello, Experimentally testable state-independent quantum contextuality. Phys. Rev. Lett. **101**(21), 210401 (2008)
19. A. Cabello, J. Estebaranz, G. García-Alcaine, Bell-Kochen-Specker theorem: a proof with 18 vectors. Phys. Lett. A **212**(4), 183–187 (1996)
20. L. Chen, D.Ž. Djoković, M. Grassl, B. Zeng, Four-qubit pure states as fermionic states. Phys. Rev. A **88**(5), 052309 (2013)

21. C. Chryssomalakos, E. Guzmán-González, E. Serrano-Ensástiga, Geometry of spin coherent states. J. Phys. A Math. Theor. **51**(16), 165202 (2018)
22. O. Chterental, D.Ž. Djoković, Normal forms and tensor ranks of pure states of four qubits (2006). arXiv preprint quant-ph/0612184
23. R. Cleve, R. Mittal, Characterization of binary constraint system games. in *International Colloquium on Automata, Languages, and Programming* (Springer, Berlin, Heidelberg, 2014), pp. 320–331
24. M.J. Duff, S. Ferrara, E_7 and the tripartite entanglement of seven qubits. Phys. Rev. D **76**(2), 025018 (2007)
25. W. Dür, G. Vidal, J.I. Cirac, Three qubits can be entangled in two inequivalent ways. Phys. Rev. A **62**(6), 062314 (2000)
26. A. Einstein, B. Podolsky, N. Rosen, Can quantum-mechanical description of physical reality be considered complete? Phys. Rev. **47**(10), 777 (1935)
27. W. Fulton, J. Harris, *Representation Theory*, vol. 129 (Springer Science & Business Media, Berlin, 1991)
28. I.M. Gelfand, M. Kapranov, A. Zelevinsky, *Discriminants, Resultants, and Multidimensional Determinants* (Springer Science and Business Media, Berlin, 2008)
29. J. Harris, *Algebraic Geometry: A First Course*, vol. 133 (Springer Science & Business Media, Berlin, 2013)
30. H. Havlicek, B. Odehnal, M. Saniga, Factor-group-generated polar spaces and (multi-) qudits. Symmetry Integr. Geom. Methods Appl. **5**, 096 (2009)
31. H. Heydari, Geometrical structure of entangled states and the secant variety. Quantum Inf. Process. **7**(1), 43–50 (2008)
32. F. Holweck, H. Jaffali, Three-qutrit entanglement and simple singularities. J. Phys. A Math. Theor. **49**(46), 465301 (2016)
33. F. Holweck, P. Lévay, Classification of multipartite systems featuring only $|W\rangle$ and $|GHZ\rangle$ genuine entangled states. J. Phys. A Math. Theor. **49**, 085201 (2016)
34. F. Holweck, M. Saniga, Contextuality with a small number of observables. Int. J. Quantum Inf. **15**(04), 1750026 (2017)
35. F. Holweck, J.G. Luque, J.Y. Thibon, Geometric descriptions of entangled states by auxiliary varieties. J. Math. Phys. **53**(10), 102203 (2012)
36. F. Holweck, J.G. Luque, J.Y. Thibon, Entanglement of four qubit systems: a geometric atlas with polynomial compass I (the finite world). J. Math. Phys. **55**(1), 012202 (2014)
37. F. Holweck, J.G. Luque, M. Planat, Singularity of type D4 arising from four-qubit systems. J. Phys. A Math. Theor. **47**(13), 135301 (2014)
38. F. Holweck, M. Saniga, P. Lévay, A notable relation between N-qubit and 2^{N-1}-qubit Pauli groups via binar $LGr(N, 2N)$. Symmetry Integr. Geom. Methods Appl. **10**, 041 (2014)
39. F. Holweck, H. Jaffali, I. Nounouh, Grover's algorithm and the secant varieties. Quantum Inf. Process. **15**(11), 4391–4413 (2016)
40. F. Holweck, J.G. Luque, J.Y. Thibon, Entanglement of four-qubit systems: a geometric atlas with polynomial compass II (the tame world). J. Math. Phys. **58**(2), 022201 (2017)
41. M. Howard, J. Wallman, V. Veitch, J. Emerson, Contextuality supplies the "magic" for quantum computation. Nature **510**(7505), 351–355 (2014)
42. V.G. Kac, V.L. Popov, E.B. Vinberg, Sur les groupes lineaires algebriques dont l'algebres des invariants est libres. CR Acad. Sci. Paris **283**, 875–878 (1976)
43. G. Kirchmair, F. Zähringer, R. Gerritsma, M. Kleinmann, O. Gühne, A. Cabello, R. Blatt, C.F. Roos, State-independent experimental test of quantum contextuality. Nature **460**(7254), 494–497 (2009)
44. S. Kochen, E.P. Specker, The problem of hidden variables in quantum mechanics, in *The Logico-Algebraic Approach to Quantum Mechanics* (Springer, Dordrecht, 1975), pp. 293–328
45. J.M. Landsberg, *Tensors: Geometry and Applications* (American Mathematical Society, Providence, 2012)
46. J.M. Landsberg, L. Manivel, The projective geometry of Freudenthal's magic square. J. Algebra **239**(2), 477–512 (2001)

47. J.M. Landsberg, L. Manivel, On the ideals of secant varieties of Segre varieties. Found. Comput. Math. **4**(4), 397–422 (2004)
48. J.M. Landsberg, G. Ottaviani, Equations for secant varieties of Veronese and other varieties. Ann. Mat. Pura Appl. **192**(4), 569–606 (2013)
49. C. Le Paige, Sur la théorie des formes binaires à plusieurs séries de variables. Bull. Acad. Roy. Sci. Belgique **2**(3), 40–53 (1881)
50. P. Lévay, Strings, black holes, the tripartite entanglement of seven qubits, and the Fano plane. Phys. Rev. D **75**(2), 024024 (2007)
51. P. Lévay, F. Holweck, Embedding qubits into fermionic Fock space: peculiarities of the four-qubit case. Phys. Rev. D **91**(12), 125029 (2015)
52. P. Lévay, Z. Szabó, Mermin pentagrams arising from Veldkamp lines for three qubits. J. Phys. A Math. Theor. **50**(9), 095201 (2017)
53. P. Lévay, P. Vrana, Three fermions with six single-particle states can be entangled in two inequivalent ways. Phys. Rev. A **78**(2), 022329 (2008)
54. P. Lévay, M. Saniga, P. Vrana, P. Pracna, Black hole entropy and finite geometry. Phys. Rev. D **79**(8), 084036 (2009)
55. P. Lévay, M. Planat, M. Saniga, Grassmannian connection between three-and four-qubit observables, Mermin's contextuality and black holes. J. High Energy Phys. **2013**(9), 37 (2013)
56. P. Lévay, F. Holweck, M. Saniga, Magic three-qubit Veldkamp line: a finite geometric underpinning for form theories of gravity and black hole entropy. Phys. Rev. D **96**, 026018 (2017)
57. J.G. Luque, J.Y. Thibon, Polynomial invariants of four qubits. Phys. Rev. A **67**(4), 042303 (2003)
58. J.G. Luque, J.Y. Thibon, Algebraic invariants of five qubits. J. Phys. A Math. Gen. **39**(2), 371 (2005)
59. N.D. Mermin, Hidden variables and the two theorems of John Bell. Rev. Mod. Phys. **65**(3), 803 (1993)
60. A. Miyake, Classification of multipartite entangled states by multidimensional determinants. Phys. Rev. A **67**(1), 012108 (2003)
61. A. Miyake, F. Verstraete, Multipartite entanglement in $2 \times 2 \times n$ quantum systems. Phys. Rev. A **69**(1), 012101 (2004)
62. A. Miyake, M. Wadati, Multipartite entanglement and hyperdeterminants. Quantum Inf. Comput. **2**(7), 540–555 (2002)
63. A.G. Nurmiev, Orbits and invariants of third-order matrices. Mat. Sb. **191**(5), 101–108 (2000)
64. L. Oeding, Set-theoretic defining equations of the tangential variety of the Segre variety. J. Pure Appl. Algebra **215**(6), 1516–1527 (2011)
65. P.G. Parfenov, Tensor products with finitely many orbits. Russ. Math. Surv. **53**(3), 635–636 (1998)
66. M. Planat, M. Saniga, Five-qubit contextuality, noise-like distribution of distances between maximal bases and finite geometry. Phys. Lett. A **376**(46), 3485–3490 (2012)
67. M. Planat, M. Saniga, F. Holweck, Distinguished three-qubit 'magicity' via automorphisms of the split Cayley hexagon. Quantum Inf. Process. **12**(7), 2535–2549 (2013)
68. A. Peres, Two simple proofs of the Kochen-Specker theorem. J. Phys. A Math. Gen. **24**(4), L175 (1991)
69. M. Saniga, M. Planat, *Multiple Qubits as Symplectic Polar Spaces of Order Two*. Advanced Studies in Theoretical Physics, vol. 1 (2007), pp. 1–4
70. M. Saniga, M. Planat, Finite geometry behind the Harvey-Chryssanthacopoulos four-qubit magic rectangle. Quantum Inf. Comput. **12**(11–12), 1011–1016 (2012)
71. M. Saniga, M. Planat, P. Pracna, H. Havlicek, The Veldkamp space of two-qubits. Symmetry Integr. Geom. Methods Appl. **3**, 075 (2007)
72. M. Saniga, H. Havlicek, F. Holweck, M. Planat, P. Pracna, Veldkamp-space aspects of a sequence of nested binary Segre varieties. Ann. Inst. Henri Poincaré D **2**(3), 309–333 (2015)
73. M. Sanz, D. Braak, E. Solano, I.L. Egusquiza, Entanglement classification with algebraic geometry. J. Phys. A Math. Theor. **50**(19), 195303 (2017)

74. G. Sárosi, P. Lévay, Entanglement in fermionic Fock space. J. Phys. A Math. Theor. **47**(11), 115304 (2014)
75. A. Sawicki, V.V. Tsanov, A link between quantum entanglement, secant varieties and sphericity. J. Phys. A Math. Theor. **46**(26), 265301 (2013)
76. A. Terracini, Sulle v_k per cui la varieta degli $s_h(h + 1)$ seganti ha dimensione minore dell'ordinario. Rendiconti del Circolo Matematico di Palermo (1884–1940), **31**(1), 392–396 (1911)
77. K. Thas, The geometry of generalized Pauli operators of N-qudit Hilbert space, and an application to MUBs. Europhys. Lett. **86**(6), 60005 (2009)
78. F. Verstraete, J. Dehaene, B. De Moor, H. Verschelde, Four qubits can be entangled in nine different ways. Phys. Rev. A **65**(5), 052112 (2002)
79. P. Vrana, P. Lévay, Special entangled quantum systems and the Freudenthal construction. J. Phys. A Math. Theor. **42**(28), 285303 (2009)
80. P. Vrana, P. Lévay, The Veldkamp space of multiple qubits. J. Phys. A Math. Theor. **43**(12), 125303 (2010)
81. M. Waegell, P.K. Aravind, Proofs of the Kochen-Specker theorem based on a system of three qubits. J. Phys. A Math. Theor. **45**(40), 405301 (2012)
82. M. Waegell, P.K. Aravind, Proofs of the Kochen-Specker theorem based on the N-qubit Pauli group. Phys. Rev. A **88**(1), 012102 (2013)
83. J. Weyman, A. Zelevinsky, Singularities of hyperdeterminants. Ann. Inst. Four. **46**(3), 591–644 (1996)
84. F.L. Zak, *Tangents and Secants of Algebraic Varieties*. Translations of Mathematical Monographs, vol. 127 (American Mathematical Society, Providence, 1993)

Chapter 6
Hilbert Functions and Tensor Analysis

Luca Chiantini

Abstract We show how well known tools of algebraic geometry for the study of finite sets can be fruitfully applied to the study of Waring decompositions of symmetric tensors (forms). We mainly focus on the uniqueness of a given decomposition (the identifiability problem), and show how, in some cases, one can effectively determine the uniqueness even in some range in which the Kruskal's criterion does not apply.

6.1 Introduction

The paper aims to introduce some basic geometric methods for the study of the decompositions of tensors. It is mainly devoted to symmetric decompositions of symmetric tensors, which can be identified with homogeneous polynomials, i.e. forms.

Decomposing a form F as a sum of powers (Waring decomposition) is a crucial step to understand the complexity of F. The complexity, or (Waring) rank, of F is indeed given by the minimal number of summands which are necessary to express F as a sum of powers.

In many effective cases, it turns out that one has one decomposition of F as a sum of powers, and the problem is to determine if the given decomposition has minimal length or it is unique (up to trivialities). Just to give a couple of examples:

– in the Strassen problem, one has a form which is a sum $F = F_1 + F_2$ where F_1, F_2 are forms defined over two different, disjoint sets of variables. Then one can assume to have a minimal decomposition of both F_1 and F_2. The problem is

L. Chiantini (✉)
Dipartimento di Ingegneria dell'Informazione e Scienze Matematiche, Università di Siena, Siena, Italy
e-mail: luca.chiantini@unisi.it

© Springer Nature Switzerland AG 2019
E. Ballico et al. (eds.), *Quantum Physics and Geometry*,
Lecture Notes of the Unione Matematica Italiana 25,
https://doi.org/10.1007/978-3-030-06122-7_6

to determine if the sum of the two decompositions gives a decomposition of F *of minimal length*. See [10] and [27], for recent accounts on the theory.

- in the application of tensor analysis to signal processing, there are computational methods which can determine (an approximation of) one decomposition of a tensor F. Since one aims to reconstruct the original components of a mixed signal, the uniqueness of the decomposition is crucial to guarantee that the computed decomposition is (in a small neighborhood of) the correct one (see e.g. [26]).

For the identifiability problem, i.e. in order to determine that a decomposition is unique (up to trivialities), the most popular criterion is the Kruskal's criterion (see Theorem 6.9 below), which requires the calculation of the Kruskal rank of a set of points (see Definition 6.1). Kruskal's criterion only works for small values of the rank. Recently, for symmetric tensors, there is a series of results which show how the Kruskal's criterion can be modified, to widen slightly the range of application (see [2, 4, 5, 14]). These extensions of Kruskal's criterion are mainly based on methods of algebraic geometry for the study of finite sets in projective spaces.

Since we believe that geometric tools for the study of finite projective sets can contribute to many other aspects of the theory of symmetric (and maybe also non-symmetric) tensors, and we feel that several tools are not widely known in the community of researchers in tensor analysis, we provide here an account of methods which constitute the background for the theory developed in [14] and [2].

As a by-product, we show how similar argument yield a slight extension of the results of [2], for forms of degree 4, even to the case in which the Kruskal rank of a given decomposition is not maximal (see Theorem 6.12).

We hope, in this way, to contribute to the propagation of geometric tools which can help a lot our insight into the analysis of decompositions of specific tensors.

The structure of the paper is the following. The first section contains some basic definitions, basic results and remarks which are useful in the theory. The second section contains a list of results on tensors which are proved by means of the Hilbert function. The third section is devoted to prove a new result, which extends a recent criterion, proved by Angelini, Vannieuwenhoven and the author [2], for the (symmetric) identifiability of a symmetric tensor in a range where the Kruskal's criterion does not apply. The result requires a deep analysis of the Hilbert function of a finite set in a projective space. In the last section there is a short list of possible developments of the theory and open problems.

6.2   Tensors and Projective Geometry

Since the study of tensors under a geometric point of view is strictly related with systems of homogeneous polynomials and their solutions, it is natural, from a mathematical point of view, to treat tensors defined over an algebraically closed field, as the complex field $\mathbb{C}$.

At the risks of losing a strict connection with experience, yet the choice of working over $\mathbb{C}$ will not sound so odd to specialists of quantum information theory, where the algebraic properties of complex numbers play a primary role in many quantum manipulations.

Less familiar is the choice of working on *projective* spaces of tensors. The idea behind using the projective setting is that the phenomena encoded in a tensor T are as well encoded in its multiples aT, for $a \in \mathbb{C}$ a non-zero constant. In projective spaces, a point P is an equivalence class containing a vector and its multiples. At the cost of dropping the one-to-one correspondence between points and coordinates (which are defined up to *scaling*), projective geometry provides a compact algebraic ambient where some operations, like linear dependence, have a natural interpretation.

Thus, we drop the *probabilistic* approach, in which the sum of some entries of the tensors are forced to be 1, since they represent the probabilities of some event, and we will freely multiply tensors by (complex) scalars. It is an ubiquitous fact that all the results that we obtain can be translated in the probabilistic language, without any loss of validity. The main, non-trivial aspect of the projective point of view is the notion of *product* of projective spaces, which does not produce a linear variety.

So, we consider a complex vector space V of dimension $n + 1$, which we will often identify with $\mathbb{C}^{n+1}$, thanks to the choice of a basis. We will think of V as the space of *linear forms* $a_0x_0 + a_1x_1 + \cdots + a_nx_n$, where $x_1, \ldots, x_n$ can be identified with the elements of the chosen basis or with variables. Consequently, the space $Sym^d(V) = Sym^d(\mathbb{C}^{n+1})$ will be identified with the space of homogeneous polynomials (forms) of degree d in the $n + 1$ variables $x_0, \ldots, x_n$.

Instead of considering directly symmetric tensors as vectors of $Sym^d(V)$, we consider the projective space $\mathbb{P}(Sym^d(V))$ and consider points T in this space. Thus T corresponds to a symmetric tensor or a form, modulo scaling. Any representative for the equivalence of class of T is a *set of coordinates* for T. As $Sym^d(V)$ has dimension $\binom{n+d}{d}$, the space $\mathbb{P}(Sym^d(V))$ has projective dimension

$$N(d, n) := \binom{n + d}{d} - 1.$$

The next step is the definition of a (non-linear) map from $\mathbb{P}(V) = \mathbb{P}^n$ to the space $\mathbb{P}(Sym^d(V)) = \mathbb{P}^{N(d,n)}$: the Veronese map.

To do that, choose an order for the monomials of degree d in $n + 1$ variables $M_0, \ldots, M_N$, $N = N(d, n)$. One of the most popular order is the lexicographic one, and we will opt for it for the rest of the paper.

Then, use the coordinates to define a map $v_{d,n}$ as follows. Let a point $P \in \mathbb{P}(V)$ have coordinates $a_0x_0 + \cdots + a_nx_n$. We will write:

$$P = [a_0x_0 + \cdots + a_nx_n].$$

We define $v_{d,n}$ by sending P to the equivalence class

$$v_{d,n}(P) = [(a_0 x_0 + \cdots + a_n x_n)^d].$$

The class $v_{d,n}(P)$ does not depend on the choice of a representative for the class P, so we get a well defined projective map. We will refer to this map as the *Veronese map of degree d in n + 1 variables*. We will often write v_d for the Veronese map, when there is no confusion on the number of variables.

We notice that the Veronese maps are embeddings.

Proposition 6.1 *Every Veronese map $v_{d,n}$ is injective.*

Proof Assume that two points $P, Q \in \mathbb{P}^n$ have the same image in $v_{d,n}$. Choose coordinates in $\mathbb{P}(V)$ and let $a_0 x_0 + \cdots + a_n x_n$ be coordinates for P and $b_0 x_0 + \cdots + b_n x_n$ be coordinates for Q. Then $(b_0 x_0 + \cdots + b_n x_n)^d$ is equal to $\alpha(a_0 x_0 + \cdots + a_n x_n)^d$, for some $\alpha \in \mathbb{C} \setminus \{0\}$. Since $\mathbb{C}$ is algebraically closed, then, after scaling $a_0 x_0 + \cdots + a_n x_n$ by a d-root of α, we may assume $(b_0 x_0 + \cdots + b_n x_n)^d = (a_0 x_0 + \cdots + a_n x_n)^d$. Thus $b_i = \epsilon_i a_i$, for some choice of the d-roots of unit ϵ_i, $i = 0, \ldots, n$. We want to prove that the ϵ_i's are all equal, so that $P = Q$. Indeed, since $\epsilon_0^{(d-j)} \epsilon_i^j = 1$ for all i, j, multiplying by ϵ_0^j it follows $\epsilon_0^j = \epsilon_i^j$ for any j, hence $\epsilon_0 = \epsilon_i$ for all i.

Notice that the previous construction is not the unique way to define a Veronese map. Often $v_{d,n}(P)$ is defined by computing $b_i = M_i(a_0, \ldots, a_n)$ for $i = 0, \ldots, N$ and sending P to the equivalence class $[b_0 M_0 + \cdots + b_N M_N]$. We made our choice in order to make it obvious that the image of the Veronese map is the set of forms which are a power of a linear forms. Since the two choices differ only by the multiplication by a non-singular diagonal matrix, the geometric properties will not be affected after taking any of the choices.

Next, we need to fix some notation for finite subsets of a projective space.

Let $A \subset \mathbb{P}^n$ be a non-empty finite set. We denote by $\ell(A)$ the cardinality of A. We will say that A is *linearly independent* when choosing a set of coordinates for each point of A we get a set of linearly independent vectors. This definition does not depend on the choice of the coordinates for each point.

We will denote with $\langle A \rangle$ the *linear span* of A.

Remark 6.1 The projective dimension of $\langle A \rangle$ is at most $\ell(A) - 1$. The dimension of $\langle A \rangle$ is equal to $\ell(A) - 1$ precisely when A is linearly independent.

Notice that, by elementary linear algebra, for any finite set $A \subset \mathbb{P}^n = \mathbb{P}(V)$ the dimension of the linear span $\langle A \rangle$ is equal to n minus the dimension of the space of linear forms that vanish at the points of A.

Definition 6.1 Let $A \subset \mathbb{P}^n$ be a finite set. The *Kruskal rank* is the maximum integer k_A such that any subset $B \subset A$ of cardinality $\ell(B) \leq k_A$ is linearly independent.

Notice that k_A is at most equal to $\ell(A)$, and $k_A = \ell(A)$ if and only if A is linearly independent. Unless A is a singleton, then k_A is always bigger than 1. Moreover $k_A = 2$ exactly when A is aligned.

Obviously the Kruskal rank of a set of points $A \subset \mathbb{P}^n$ cannot exceed neither $n + 1$, nor the cardinality of A. We have indeed:

$$k_A \leq \dim\langle A\rangle + 1 \leq \ell(A).$$

Next definition concerns the case where the Kruskal rank is maximal.

Definition 6.2 A finite set $A \subset \mathbb{P}^n$ is in *linear general position* (LGP) if the Kruskal rank of A is maximal, i.e. the Kruskal rank is equal to $\min\{\ell(A), n + 1\}$. This is equivalent to say that for any $a \leq n + 1$, any subset of A of cardinality a is linearly independent.

Next, we come to the definition of *decomposition* of a (symmetric) tensor.

Definition 6.3 Let $A \subset \mathbb{P}^n = \mathbb{P}(V)$ be a finite set. We say that A is a *decomposition* of the tensor $T \in \mathbb{P}(Sym^d(V))$, or equivalently that A *computes* T, if T belongs to the span $\langle v_d(A)\rangle$.

Definition 6.4 Let $A \subset \mathbb{P}^n$ be a decomposition of T. A is *minimal* if we cannot find a proper subset A' of A such that $T \in \langle v_d(A')\rangle$.

Remark 6.2 If $A \subset \mathbb{P}^n$ is a decomposition of T and satisfies the minimality property, then in particular the points of $v_d(A)$ are linearly independent, i.e.,

$$\dim(\langle v_d(A)\rangle) = \ell(A) - 1.$$

6.2.1   The Hilbert Function of Finite Sets in Projective Spaces

We collect in this section a series of definitions and propositions which are well known to people working in algebraic geometry, but maybe not so familiar to other people working in tensor analysis. The main definition is the *Hilbert function* of a finite set in a projective space, which is a basic tool for our results on the decompositions of symmetric tensors.

Definition 6.5 Let $Y \subset \mathbb{C}^{n+1}$ be an ordered, finite set of cardinality ℓ of vectors. Fix an integer $d \in \mathbb{N}$.

The *evaluation map of degree d on Y* is the linear map

$$ev_Y(d) : Sym^d(\mathbb{C}^{n+1}) \to \mathbb{C}^\ell$$

which sends $F \in Sym^d(\mathbb{C}^{n+1})$ to the evaluation of F at the vectors of Y.

We will use the evaluation map to define the Hilbert function of a finite set $Z \subset \mathbb{P}^n$.

Remark 6.3 Let $A \subset \mathbb{P}^n$ be a finite set, with a definite order. Choose a set of homogeneous coordinates for the points of A. We get an ordered set of vectors $Y \subset \mathbb{C}^{n+1}$, for which the evaluation map $ev_Y(d)$ is defined for every d.

If we change the choice of the homogeneous coordinates for the points of the fixed set A, we get another ordered set $Y' \subset \mathbb{C}^{n+1}$ and the evaluation map $ev_{Y'}(j)$ differs from $ev_Y(j)$ for the multiplication by a non-singular diagonal matrix. Thus the rank of $ev_Y(j)$ and $ev_{Y'}(j)$ are the same for all j.

It is also clear that the rank of $ev_Y(j)$ does not depend on how we ordered the points of A.

Let $f : \mathbb{C}^{n+1} \to \mathbb{C}^{n+1}$ be an automorphism and consider the associated change of coordinates $\mathbb{P}^n \to \mathbb{P}^n$, that we call again f, by abuse. Then the evaluation on Y and $f(Y)$ differ by the multiplication by a non-singular matrix. Thus for any d the maps $ev_Y(d)$ and $ev_{f(Y)}(d)$ have the same rank.

Definition 6.6 Let $Z \subset \mathbb{P}^n$ be a finite set. Choose an order and an ordered set of homogeneous coordinates Y for the points of A. Define the *Hilbert function* of Z as the map

$$h_Z : \mathbb{Z} \to \mathbb{N} \qquad h_Z(d) = \mathrm{rank}(ev_Y(d)).$$

By the previous remark, the Hilbert function does not depend on the choice of the coordinates, as well as it does not vary after a change of coordinates in $\mathbb{P}^n$.

People who are expert of algebraic geometry may wonder why we did not define the Hilbert function as the rank of the restriction maps $H^0(\mathcal{O}(d)) \to H^0(\mathcal{O}_Z(d))$, where $\mathcal{O}$, $\mathcal{O}_Z$ indicate respectively the structure sheaves of $\mathbb{P}^n$ and A. This would simplify the notation, since the restriction is well defined, regardless of a choice of coordinates for the points of A. On the other hand, our definition is immediately accessible also to readers who are not expert about cohomology, structure sheaves and so on. We preferred to make our basic definition more familiar and easily computable for a wider audience. We based our definition on the choice of coordinates because only after a choice of coordinates for the points of A one has a natural identification of $H^0(\mathcal{O}_Z(d))$ with $\mathbb{C}^{\ell}$.

There is a different notation for the Hilbert function, which is widely used in algebraic geometry. Since it clarifies some aspects, we introduce it.

Remark 6.4 Recall that the homogeneous ideal I_Z of the set Z in the polynomial ring $\mathbb{C}[t_0, \ldots, t_n]$ is the ideal generated by all the homogeneous polynomials (forms) which vanish at all the points of Z. Thus, I_Z is a graded ideal. Its degree d summand $I_Z(d)$ is exactly the kernel of the evaluation map $ev_Z(d)$.

Notice that, indeed, the kernel does not depend on the choice of homogeneous coordinates for the points of Z, because the vanishing of a form at a projective point P is independent from the choice of a specific set of homogeneous coordinates for P.

Thus, recalling that the vector space of forms of degree d we have

$$h_Z(d) = \dim(Sym^d(\mathbb{C}^{n+1})) - \dim(I_Z(d)) = \binom{n+d}{n} - \dim(I_Z(d)).$$

Consequently, we introduce the following notation:

Definition 6.7 Let Z be a finite subset of the projective space $\mathbb{P}^n$ and let h_Z be its Hilbert function. For any $d \geq 0$, the value $h_Z(d)$ is also called the *number of conditions that Z imposes to forms of degree d*.

We say that Z *imposes independent conditions to forms of degree d*, or also that *the points of Z are separated by forms of degree d*, if $h_Z(j) = \ell(Z)$. This happens exactly when, for (any choice of) a set Y of homogeneous coordinates for the points of Z, the evaluation map $ev_Y(d)$ surjects.

Remark 6.5 Let us explain in more details the last definition. Set $\ell = \ell(Z)$, and fix an order for the points of Z.

Take a vector $e_j = (0, \ldots, 0, 1, 0, \ldots, 0)$ (1 is in the j-th position) of the natural basis of $\mathbb{C}^\ell$, which corresponds to the j-th point P_j of Z in the given order. We say that P_j *is separated in Z by forms of degree d* if e_j belongs to the image of the evaluation map $ev_Y(d)$. Indeed, in this case, e_j is the evaluation of a form F of degree d. Thus there exists a form F which vanishes at all the points of Z, but P_j. Notice that this is independent on the choice of the homogeneous coordinates Y.

If $h_Z(j) = \ell(Z)$, i.e. if the evaluation map $ev_Y(d)$ surjects, then any point of Z is separated.

The link between the Hilbert function of finite sets and the decompositions of symmetric tensors is mainly based on the following formula, which gives a different, geometric interpretation of the values $h_Z(d)$.

Proposition 6.2 *Let $v_{d,n} : \mathbb{P}^n \to \mathbb{P}^N$, $N = N(d, n)$, be the d-th Veronese embedding of $\mathbb{P}^n$. For any finite set $Z \subset \mathbb{P}^n$, and for any $d \geq 0$, the value $h_Z(d)$ determines the dimension of the span of $v_d(Z)$. I.e.:*

$$h_Z(d) = \dim(\langle v_{d,n}(Z) \rangle) + 1.$$

Proof We know that the value $h_Z(d)$ is equal to the dimension of $Sym^d(\mathbb{C}^{n+1})$ minus the dimension of the space $I_Z(d)$, where I_Z is the homogeneous ideal of Z in $\mathbb{C}[t_0, \ldots, t_n]$. If we identify the coordinates in $\mathbb{P}^{N(d,n)} = \mathbb{P}(Sym^d(\mathbb{C}^{n+1}))$ with the monic monomials M_j's of degree d in $\mathbb{C}[t_0, \ldots, t_n]$ (say with the lexicographic order), then any element of $I_Z(d)$ corresponds to a linear form in $\mathbb{P}(Sym^d(\mathbb{C}^{n+1}))$. The claim follows by Remark 6.1.

Definition 6.8 We define the *first difference of the Hilbert function* Dh_Z of Z as:

$$Dh_Z(j) = h_Z(j) - h_Z(j-1), \quad j \in \mathbb{Z}.$$

The set of non-zero values of Dh_Z is called the *h-vector* of Z.

The following properties of h_A and Dh_A are elementary and well-known in algebraic geometry. We recall them because they will be useful throughout the paper.

Lemma 6.1 *Set $\ell = \ell(Z)$. Then we have:*

(i) $h_Z(d) \le \ell$ for all d;
(ii) $Dh_Z(d) = 0$ for $d < 0$;
(iii) $h_Z(0) = Dh_Z(0) = 1$;
(iv) $Dh_Z(d) \ge 0$ for all d;
(v) $h_Z(d) = \ell(Z)$ for all $d \ge \ell(Z) - 1$;
(vi) $h_Z(i) = \sum_{0 \le d \le i} Dh_Z(d)$;
(vii) $Dh_Z(d) = 0$ for $d \gg 0$ and $\sum_d Dh_Z(d) = \ell(Z)$;
(viii) if $h_Z(d) = \ell(Z)$, then $Dh_Z(d + 1) = 0$.

Proof (i) is a consequence of the definition. (ii) follows immediately since the space $Sym^d(\mathbb{C}^{n+1})$ is (0) for d negative. (iii) follows since $Sym^0(\mathbb{C}^{n+1}) = \mathbb{C}$ and the evaluation of a constant form c is equal to $c(1, \ldots, 1) \in \mathbb{C}^\ell$.

To see (iv), fix an ordered set of coordinates Y for the points of Z and fix a linear form Λ which does not vanish at any vector of the finite set Y. Then for any form F of degree d, the evaluation of ΛF at Y is equal to the evaluation of F at Y multiplied by a fixed non-singular diagonal matrix, whose entries are the evaluations of Λ at the vectors of Y. Thus the image of $ev_Y(d + 1)$ contains a subspace isomorphic to the image of $ev_Y(d)$. It follows that $h_Z(d + 1) \ge h_Z(d)$, hence $Dh_Z(d) \ge 0$.

To see (v), choose for each point $P_j \in Z$ a linear form L_j which vanishes at P_j and does not vanish at any other point $P_k \in Z$. Then for any j call F_j the product of the linear forms $L_k, k \ne j$. F_j is a form of degree $\ell - 1$, which vanishes at all the points of Z, except P_j. Thus, the evaluation of F_j at an ordered set of coordinates Y for the points of Z is a vector $(c_1, \ldots, c_\ell)$ with $c_k = 0$ for $k \ne j$ and $c_j \ne 0$. It follows that $ev_Y(\ell - 1)$ is surjective. Then, by (iv), $ev_Y(d)$ surjects for all $d \ge \ell - 1$.

(vi) is a triviality. (vii) and (viii) are obvious consequences of (v) and (vi). $\qquad \square$

Proposition 6.3 *With the previous notation, if $Z' \subset Z$, then, for every $d \in \mathbb{Z}$, we have $h_{Z'}(d) \le h_Z(d)$ and $Dh_{Z'}(d) \le Dh_Z(d)$.*

Proof Fix, as usual, an ordered set of coordinates Y', Y for the points of Z', Z respectively. Then we have an obvious forgetful map $f : \mathbb{C}^\ell \to \mathbb{C}^{\ell'}$, where $\ell' = \ell(Z')$, such that $ev_{Y'}(d) = f \circ ev_Y(d)$ for all d. This implies that $h_{Z'}(d) \le h_Z(d)$.

The second inequality is less trivial, and we will need some algebra. Write R for the polynomial ring $\mathbb{C}[t_0, \ldots, t_n]$ and call I_Z the ideal generated by forms which vanish at the points of Z. The inclusion $I_Y \subset R$ determines, for every $d \in \mathbb{Z}$ an exact sequence of vector spaces:

$$0 \to I_Y(d) \to R(d) \to (R/I)(d) \to 0,$$

where $R(d)$, $R/I_Z(d)$ are the graded pieces of the rings R, R/I respectively, in degree d. It follows by Remark 6.3 that for any d:

$$h_Z(d) = \dim(R/I_Z(d)).$$

The natural inclusion $I_Z \subset I_{Z'}$ induces a surjection $R/I_Z(d) \to R/I_{Z'}(d)$ for all d. Let Λ be a linear form in $\mathbb{C}[t_0, \dots, t_n]$, which does not vanish at any point of Z. The multiplication by Λ induces an inclusion $R/I_Z(d) \to R/I_Z(d+1)$. Indeed if $F \in R(d)$ is a form which does not vanish at some point $P \in Z$, then LF cannot vanish at P, i.e. the class of LF is non-zero in $R/I_Z(d+1)$. Call J_Z the ideal generated by I_Z and Λ. We have an exact sequence:

$$0 \to R/I_Z(d) \to R/I_Z(d+1) \to R/J_Z(d+1) \to 0$$

which proves that

$$Dh_Z(d) = \dim(R/J_Z(d+1)).$$

Similarly Λ induces an embedding $R/I_{Z'}(d) \to R/I_{Z'}(d+1)$ and $Dh_{Z'}(d) = \dim(R/J_{Z'}(d))$. Now look at the commutative diagram:

$$
\begin{array}{ccccccccc}
0 \to & R/I_Z(d) & \xrightarrow{L} & R/I_Z(d+1) & \to & R/J_Z(d+1) & \to & 0 \\
 & \downarrow & & \downarrow & & \downarrow & & \\
0 \to & R/I_{Z'}(d) & \xrightarrow{L} & R/I_{Z'}(d+1) & \to & R/J_{Z'}(d+1) & \to & 0
\end{array}
$$

Since the central vertical map $R/I_Z(d+1) \to R/I_{Z'}(d+1)$ surjects, by the snake's lemma also the map $R/J_Z(d+1) \to R/J_{Z'}(d+1)$ surjects. Then $Dh_Z(d) = \dim(R/J_Z(d+1)) \geq \dim(R/J_{Z'}(d+1)) = Dh_{Z'}(d)$. This proves the second claim.

Perhaps, the most important algebraic result on Hilbert functions of finite sets is the *maximal growth principle* found by Macaulay. Roughly speaking, the maximal growth principle gives an upper bound for the value of $h_A(i+1)$ in terms of $h_Z(i)$ and the dimension of the ambient space. We list below the most relevant consequences for the application to the study of tensors and forms.

Proposition 6.4 *Assume that for some $j > 0$ we have $Dh_Z(j) \leq j$. Then:*

$$Dh_Z(j) \geq Dh_Z(j+1).$$

In particular, if for some $j > 0$, $Dh_Z(j) = 0$, then $Dh_Z(i) = 0$ for all $i \geq j$.

Proof See Section 3 of [8].

Example 6.1 Let us see what happens for $h_Z(1)$. Since for $i = 1$ the domain of the evaluation map is $Sym^1(\mathbb{C}^{n+1}) = \mathbb{C}^{n+1}$, then clearly $h_Z(1) \leq n+1$. So $h_Z(1) = 0$

can hold only if $\ell(Z) \leq n + 1$. Moreover the kernel of the evaluation map $ev_Z(1)$ is isomorphic to the space of linear forms in $\mathbb{P}^n$ which vanish at Z. Thus:

$$h_Z(1) = 1 + \dim(\langle Z \rangle).$$

In particular, $h_Z(1) = 0$ if and only if Z is linearly independent.

Remark 6.6 Assume that for some j we have $Dh_Z(j) = 0$, so that $h_Z(j-1) = h_Z(j)$. By Proposition 6.4, for any $i \geq j$ also $Dh_Z(i) = 0$, i.e., $h_Z(j-1) = h_Z(i)$ for any $i \geq j$. Therefore, by part (v) of Lemma 6.1, $h_Z(j-1)$ is equal to the cardinality of Z, i.e., the evaluation map in degree $j-1$ surjects and Z imposes independent conditions to hypersurfaces of degree $j-1$.

Remark 6.7 Assume $h_Z(i) = \ell(Z) - 1$. Then $h_Z(i+1) > h_Z(i)$, by Remark 6.6. Thus, if $h_Z(i) = \ell(Z) - 1$, then necessarily $h_Z(i+1) = \ell(Z)$.

Hilbert functions of finite sets share many other properties. One can find an accurate account of the theory in the book of Iarrobino and Kanev [20] and in the book of Migliore [24].

We will focus on the *Cayley-Bacharach* property, which is defined as follows:

Definition 6.9 A finite set $Z \subset \mathbb{P}^n$ satisfies the *Cayley-Bacharach property in degree i*, abbreviated as $CB(i)$, if, for any $P \in Z$, every form of degree i vanishing at $Z \setminus \{P\}$ also vanishes at P.

Remark 6.8 One should compare CB with the property of separating points. In a sort of sense, the CB property is the contrary of the separation property.

- Z is separated in degree i if for all $P \in Z$, there exists a form of degree i vanishing at $Z \setminus \{P\}$ and not vanishing at P.
- Z does not satisfy CB if there exists $P \in Z$ and there exists a form of degree i vanishing at $Z \setminus \{P\}$ and not vanishing at P.

In particular, if Z satisfies $CB(i)$, then hypersurfaces of degree i cannot separate the points of Z, i.e. $h_Z(i) < \ell(Z)$.

Example 6.2 The set Z consisting of four points in $\mathbb{P}^2$, three of them aligned, does not satisfy $CB(1)$, and $h_Z(1) < 4$.

Let Z be a set of 6 points in $\mathbb{P}^2$.

If the 6 points are general, then $Dh_Z = (1, 2, 3)$, and Z satisfies $CB(1)$. Since $h_Z(2) = 6$, Z does not satisfy $CB(2)$.

If Z lies on an irreducible conic, then $Dh_Z = (1, 2, 2, 1)$, and Z satisfies $CB(2)$, and, hence, $CB(1)$.

If Z has 5 points on a line plus one point off the line, then $Dh_Z = (1, 2, 1, 1, 1)$, and Z does not satisfy $CB(1)$.

Remark 6.9 If Z satisfies $CB(i)$, then it satisfies $CB(i-1)$ too. Otherwise, one could find $P \in Z$ and a hypersurface $F \subset \mathbb{P}^n$ of degree $(i-1)$ such that $Z \setminus \{P\} \subset F$ and

$P \notin F$. Therefore, if $H_P \subset \mathbb{P}^n$ is a hyperplane not containing P, then $F \cup H_P \in H^0(J_{Z\setminus\{P\}}(i)) \setminus H^0(J_Z(i))$, which contradicts the hypothesis.

Remark 6.10 Assume that Z satisfies $CB(i)$. Call I_Z the homogeneous ideal of Z. For any $P \in Z$ call $I_{Z\setminus\{P\}}$ the homogeneous ideal of $Z \setminus \{P\}$. Then for all $j \leq i$ we have I_Z and $I_{Z\setminus\{P\}}$ are equal in degree j. It follows that:

$$h_Z(j) = h_{Z\setminus\{P\}}(j) \quad \text{and} \quad Dh_Z(j) = Dh_{Z\setminus\{P\}}(j) \quad \forall j \leq i. \tag{6.1}$$

The following proposition, which gives a strong bound on the Hilbert function of sets with a Cayley-Bacharach property, is a refinement of a result due to Geramita, Kreuzer, and Robbiano (see Corollary 3.7 part (b) and (c) of [18]).

Theorem 6.1 *If a finite set $Z \subset \mathbb{P}^n$ satisfies $CB(i)$, then for any j such that $0 \leq j \leq i+1$ we have*

$$Dh_Z(0) + Dh_Z(1) + \cdots + Dh_Z(j) \leq Dh_Z(i+1-j) + \cdots + Dh_Z(i+1).$$

Proof See Theorem 4.9 of [2].

Finally, let us point out the relation between the Hilbert functions of a finite set Z and of its image in a Veronese map $v_d(Z)$.

Remark 6.11 Let $Z \subset \mathbb{P}^n$ be a finite set and let $v_d(Z) \subset \mathbb{P}^N$ be its image in the d-th Veronese map. Then

$$h_Z(d) = h_{v_d(Z)}(1).$$

Namely the inverse image in v_d of a linear form Λ in $\mathbb{P}^N$ corresponds to a form of degree d in $\mathbb{P}^n$, and the consequent map $\mathbb{C}^{N+1} \to Sym^d(\mathbb{C}^{n+1})$ surjects. Moreover it is easy to see that, for any choice of coordinates Y for the points of Z in $\mathbb{P}^n$ and the consequent choice $v_d(Y)$ of coordinates for the points of $v_d(Z)$, one has $ev_{Y'}(L) = ev_Y(v_d^{-1}(L))$, so that the claim follows.

In particular, since v_d is a bijection, then $h_Z(d) = \ell(Z)$ if and only if $h_{v_d(Z)}(1) = \ell(v_d(Z))$, i.e. if and only if $v_d(Z)$ is linearly independent (see Example 6.1).

The following result will be useful in the proof of Theorem 6.12

Proposition 6.5 *Let Z be a finite set in $\mathbb{P}^n$. Call k the Kruskal rank of Z. If $\ell(Z) \leq 2k - 1$, then Z is separated by forms of degree 2. Hence $v_2(Z)$ is linearly independent.*

Proof We know that $k \leq n + 1$. For any point $P \in Z$, consider a partition of the residue $Z \setminus \{P\}$ in two disjoint sets Z_1, Z_2, each of cardinality at most $k - 1$. Since $k - 1 \leq n$, then the span L_i of Z_i has dimension strictly smaller than n. Moreover, L_i does not contain P, for otherwise Z has k linearly dependent points, which contradicts the assumption on the Kruskal rank of Z. Thus, there are hyperplanes H_1, H_2 containing Z_1 and Z_2 respectively and both missing P. The union $Q = H_1 \cup H_2$ is a quadric which misses P and contains the remaining points of Z.

6.3   Results on Tensors from Classical Projective Geometry

The section is devoted to list a series of results on tensors whose proof is based on the study of the Hilbert function of finite sets. In many cases we omit the proof, or give only a short draft it.

Remark 6.12 Fix integers $d, n > 1$ and consider symmetric tensors in the space $\mathbb{P}(Sym^d(\mathbb{C}^{n+1}))$. In [1] Alexander and Hirschowitz determined the unique value $r_{d,n}$ such that the set of tensors of rank $r_{d,n}$ is dense in $\mathbb{P}(Sym^d(\mathbb{C}^{n+1}))$. It turns out that $r_{d,n}$ coincides with the expected value, except for a short list of exceptions.

We will call $r_{d,n}$ the *generic rank*.

Definition 6.10 We say that a tensor $T \in \mathbb{P}(Sym^d(\mathbb{C}^{n+1}))$ of rank r is *identifiable* if T has only one minimal decomposition A with $\ell(A) = r$, up to scaling and permutations of the summands.

Identifiability is a relevant property for tensors for many applications, as explained in Sect. 6.1.

If we fix a *subgeneric* value of the rank $r < r_{d,n}$, then the set of tensors of rank $\leq r$ in $\mathbb{P}(Sym^d(\mathbb{C}^{n+1}))$ is irreducible and its general element has rank r, so we can talk about a *general* tensor of rank r. For general tensors of rank $r < r_{d,n}$, thanks to the fundamental preparatory works [1, 11], and [3], the situation with respect to the identifiability property has been completely described in [13].

Theorem 6.2 *Let $d, r \geq 2$. The general tensor in $\mathbb{P}(Sym^d(\mathbb{C}^{n+1}))$ of subgeneric rank $r < r_{d,n}$ is identifiable, unless it is one of the following cases:*

1. $d = 2$;
2. $d = 6$, $n = 2$, and $r = 9$;
3. $d = 4$, $n = 3$, and $r = 8$;
4. $d = 3$, $n = 5$, and $r = 9$.

In the first case there are infinitely many decompositions. In the three last exceptional cases, there are exactly two decompositions.

Proof See Theorem 1.1 of [13]. $\qquad \square$

Remark 6.13 On the contrary, when $r = r_{d,n}$, there are very few cases in which a general tensor of rank r is identifiable. The classification has been proved by Galuppi and Mella, see [17].

When $r > r_{d,n}$, the situation is less known. It is not even obvious what is the meaning of *generic tensors*, since the set of tensors of given rank can have many components.

In any case, one expects that a sufficiently general tensor is *not* identifiable, though for $r > r_{n,d}$ very few things are known.

For the case $r = r_{n,d}$, the situation is completely described in [1, 23] and mainly in [17]: there are many decompositions, unless d, n are included in a short list of cases.

Let us turn to the problem of the identifiability of one *specific* given tensor $T \in \mathbb{P}(Sym^d(\mathbb{C}^{n+1}))$, of which we know a minimal decomposition $A \subset \mathbb{P}^n = \mathbb{P}(\mathbb{C}^{n+1})$ with $\ell(A) = r$.

Recall that *minimal* means that the set $v_d(A)$ is linearly independent. We *do not* assume that $\ell(A)$ is actually the rank of T, i.e. we do not know if T has some other decomposition with smaller cardinality.

Let us start recalling the following, classical result of Sylvester, which disposes of the case $n = 1$, the case of *binary forms*:

Theorem 6.3 *Assume $n = 1$, i.e. consider the space of tensor $\mathbb{P}(Sym^d(\mathbb{C}^2))$. Then $r_{2,d} = (d+1)/2$ if d is odd, $r_{2,d} = (d+2)/2$ if d is even. Moreover every tensor of rank $r < r_{2,d}$ is identifiable.*

Proof See [28].

Indeed, to be precise, when $n = 1$ and d is odd, also tensors of rank $r_{2,d}$ are identifiable. See Theorem 6.4 below.

So, we restrict ourselves to the case $n > 1$.

The reason why an analysis of the Hilbert functions is relevant for the identifiability property is expressed in the following lemma, which can be found in [4]:

Lemma 6.2 *Consider two different minimal decompositions A, B of a tensor $T \in \mathbb{P}(Sym^d(\mathbb{C}^{n+1}))$. In other words, we have:*

$$T \in \langle v_d(A) \rangle \cap \langle v_d(B) \rangle.$$

Then if $Z = A \cup B$, we get $h_Z(d) < \ell(Z)$, so that $Dh_Z(d+1) > 0$.

Proof Set $Z = A \cup B$. First assume that A, B are disjoint. The existence of T implies that $v_d(Z)$ is not linearly independent. By Example 6.1, this implies that linear forms in the space $\mathbb{P}^N$ spanned by $v_d(\mathbb{P}^n)$ do not separate the points of $v_d(Z)$. By Remark 6.11, this implies that forms of degree d in $\mathbb{P}^n$ do not separate the points of Z. The claims follow by part viii) of Lemma 6.1 and Proposition 6.4.

If $A \cap B \neq \emptyset$, define $B' = A \setminus B$, so that Z is the disjoint union of A and B'. By elementary linear algebra, $\langle v_d(A) \rangle \cap \langle v_d(B) \rangle$ is also spanned by $v_d(A \cap B)$ and $\langle v_d(A) \rangle \cap \langle v_d(B') \rangle$. By the minimality of A, T cannot belong to the span of $v_d(A \cap B)$. Thus $\langle v_d(A) \rangle \cap \langle v_d(B') \rangle$ is non empty, and the claim follows again, as above, by part viii) of Lemma 6.1 and Proposition 6.4.

We can be more precise about the dimension of the intersection of the span of $v_d(A)$ and $v_d(B)$.

Lemma 6.3 *Let A, $B \subset \mathbb{P}^n$ be disjoint finite sets. Set $Z = A \cup B$. Then:*

$$\dim(\langle v_d(A) \rangle \cap \langle v_d(B) \rangle) = \ell(Z) - h_Z(d) - 1.$$

If $A \cap B \neq \emptyset$, then:

$$\dim(\langle v_d(A) \rangle \cap \langle v_d(B) \rangle) \leq \dim(v_d(A \cap B)) + \ell(Z) - h_Z(d).$$

Proof The first formula in an exercise for the application of the Grassmann intersection formula. The second formula follows since, setting $B_0 = B \setminus A$ so that A, B_0 are disjoint and $Z = A \cup B_0$, by elementary linear algebra $\langle v_d(A) \rangle \cap \langle v_d(B) \rangle$ is spanned by $v_d(A \cap B)$ and $\langle v_d(A) \rangle \cap \langle v_d(B_0) \rangle$.

An extension of Sylvester's theorem, which works for *all* symmetric tensors in $\mathbb{P}(Sym^d(\mathbb{C}^{n+1}))$, is possible for $n > 1$ only for small values of the rank. The following statement is proved in Theorem 1.5.1 of [9]. We give here an alternative proof, in terms of the Hilbert function of decompositions.

Theorem 6.4 *Assume that a tensor $T \in \mathbb{P}(Sym^d(\mathbb{C}^{n+1}))$ has a decomposition A with $\ell(A) \leq (d + 1)/2$. Then T has rank $\ell(A)$ and it is identifiable.*

Proof Assume on the contrary that T has a second decomposition B with $\ell(B) \leq \ell(A)$, and take the union $Z = A \cup B$. Then $\ell(Z) \leq 2\ell(A) \leq d + 1$. By Lemma 6.2 we have $Dh_Z(d + 1) > 0$. Thus by Proposition 6.4 and by point iii) of Lemma 6.1 we get $Dh_Z(j) > 0$ for $j = 0, \ldots, d + 1$. Hence $\sum_j Dh_Z(j) \geq d + 2$, which contradicts point vii) of Lemma 6.1.

An easy extension of Theorem 6.4 is given by the following result.

Theorem 6.5 *Assume that a tensor $T \in \mathbb{P}(Sym^d(\mathbb{C}^{n+1}))$ has a decomposition A with $\ell(A) \leq (d + n)/2$, such that $\langle A \rangle = \mathbb{P}^n$. Then T has rank $\ell(A)$ and it is identifiable.*

Proof Assume on the contrary that T has a second decomposition B with $\ell(B) \leq \ell(A)$, and take the union $Z = A \cup B$. Then $\ell(Z) \leq 2\ell(A) \leq d + n$. By Lemma 6.2 we have $Dh_Z(d + 1) > 0$. Thus by Proposition 6.4 and by point iii) of Lemma 6.1 we get $Dh_Z(j) > 0$ for $j = 0, \ldots, d + 1$. By Example 6.6 and by Proposition 6.3 we get $h_Z(1) = n + 1$, so that $Dh_Z(1) = n$. Hence $\sum_j Dh_Z(j) \geq d + n + 1$, which contradicts point vii) of Lemma 6.1.

A tensor $T \in \mathbb{P}(Sym^d(\mathbb{C}^{n+1}))$ is *concise* if there exist no linear subspaces $W \subset \mathbb{C}^{n+1}$, of codimension 1, such that T belongs to $\mathbb{P}(Sym^d(W))$.

The previous statement implies that when T is concise and it has a decomposition of cardinality $\leq (d + n)/2$, then T is identifiable.

To go further, we may assume some restrictions on the geometry of a decomposition A of T.

Lemma 6.4 *Let $Z \subset \mathbb{P}^n$ be a finite set and assume that for some $j \geq 1$: $Dh_Z(j + 1) = Dh_Z(j) = 1$. Then Z contains an aligned subset Z' of cardinality $\ell(Z') = j + 2$, and $Dh_Z(i) = Dh_{Z'}(i)$ for all $i \geq j$.*

Proof See Lemma 2 of [7]. $\square$

The following result gives a further extension of Theorem 6.4 (compare with Theorem 2 of [4]).

Proposition 6.6 *Fix a form $T \in \mathbb{P}(Sym^d(\mathbb{C}^{n+1}))$ and a minimal decomposition $A \subset \mathbb{P}^n$ of T. Assume that $\ell(A) \leq d$ and A does not contain an aligned subset of cardinality $d/2$. Then T has rank $\ell(A)$ and it is identifiable.*

Proof Assume there exists another decomposition B of T with $\ell(B) \leq d$ and call Z the union $Z = A \cup B$. Then $\ell(Z) \leq 2d$, moreover, by Lemma 6.2, $Dh_Z(d+1) > 0$, which implies $Dh_Z(d) > 0$. By Example 6.1 we get that $h_A(1) = 2$, hence also $h_Z(1) = 2$, by Now assume $Dh_Z(d) \geq 2$. Then $Dh_Z(j) \geq 2$ for $j = 1, \ldots, d$, by Proposition 6.4, so that $\sum_j Dh_Z(j) \geq 2d + 2$, which contradicts point vii) of Lemma 6.1. Then for some $j \geq 1$, $j \leq d$, we have $Dh_Z(j) < 2$. By Proposition 6.4 again, this implies $Dh_Z(d) = Dh_Z(d + 1) = 1$. Hence by Lemma 6.4, Z contains an aligned subset Z' with $\ell(Z') \geq d + 2$, and $Dh_Z(i) = Dh_{Z'}(i)$ for $i > d$. Since Z' cannot contain A, then there exists a proper subset $A' \subset A$ and a subset $B' \subset B$ such that $Z' = A' \cup B'$. Shrinking B', if necessary, we may assume that $B' \cap A = \emptyset$, so that also $A' \cap B' = \emptyset$. Then by (6.3):

$$\dim(\langle v_d(A')\rangle \cap \langle v_d(B')\rangle) = \ell(Z') - h_{Z'}(d) - 1 = \sum_{i>d} h_{Z'}(i) =$$

$$= \sum_{i>d} h_Z(i) = \dim(\langle v_d(A)\rangle \cap \langle v_d(B_0)\rangle),$$

where $B_0 = B \setminus A$. Thus:

$$\langle v_d(A')\rangle \cap \langle v_d(B')\rangle = \langle v_d(A)\rangle \cap \langle v_d(B_0)\rangle$$

Since, as in the proof of Lemma 6.3, the intersection $\langle v_d(A)\rangle \cap \langle v_d(B)\rangle$ is spanned by $v_d(A \cap B)$ and $\langle v_d(A)\rangle \cap \langle v_d(B_0)\rangle$, it follows that T belongs to the span of $v_d((A \cap B) \cup A')$. The minimality of A implies $A = (A \cap B) \cup A'$, so the points of A which are not contained in B are aligned. By assumption $\ell(A') \leq d/2$ and $\ell(A') + \ell(B') = \ell(Z') \geq d + 2$, it follows that $\ell(B') \geq 2 + d/2$. Thus $\ell(A \cap B) \leq \ell(B) - 2 - d/2 \leq \ell(A) - 2 - d/2$. Then

$$\ell(A) \leq \ell(A') + \ell(A \cap B) \leq d/2 + \ell(A) - 2 - d/2 = \ell(A) - 2,$$

a contradiction. $\square$

In order to go further in the study of the identifiability of symmetric tensors, one needs a refinement of Lemma 6.4. The refinement is provided by the following, strong result of Bigatti, Geramita and Migliore (for the case $n = 2$ the result has been proved by Davis).

Theorem 6.6 *Let $Z \subset \mathbb{P}^n$ be a finite set. Assume that for some $s \leq j$, $Dh_Z(j) = Dh_Z(j+1) = s$. Then there exists a reduced curve C of degree s such that, setting $Z' = Z \cap C$ and $Z'' = Z \setminus Z'$:*

1. *for $i \geq j - 1$, $h_{Z'}(i) = h_Z(i) - \ell(Z'')$;*
2. *for $i \leq j$, $h_Z(i) = h_C(i)$;*
3. $Dh_{Z'}(i) = \begin{cases} Dh_C(i) \text{ for } i \leq j+1; \\ Dh_Z(i) \text{ for } i \geq j. \end{cases}$

In particular, $Dh_{Z'}(i) = s$ for $s \leq i \leq j+1$.
For $n = 2$, i.e. when $Z \subset \mathbb{P}^2$, we also have:

$$h_{Z''}(j-1) = \ell(Z'') \quad and \quad Dh_{Z''}(i) = Dh_Z(i+s) - s \text{ for } i + s \leq j.$$

Proof See Theorem 3.6 of [8], and [15] for the case $n = 2$.

Thanks to Theorem 6.6, for the case $n = 2$ one can prove an extension of Proposition 6.6:

Theorem 6.7 *Fix a form $T \in \mathbb{P}(Sym^d(\mathbb{C}^3))$ and a minimal decomposition $A \subset \mathbb{P}^n$ of T. Assume that for all j the Kruskal rank of $v_j(A)$ is maximal, i.e. it is equal to the minimum between $\ell(A)$ and $\binom{j+2}{2}$. If*

$$\ell(A) < \frac{d^2 + d}{8},$$

then T has rank $\ell(A)$ and it is identifiable.

Proof See Theorem 1.4 of [5], in which the general uniform position (GUP) assumption is equivalent to the condition that the Kruskal rank of $v_j(A)$ is maximal for all j.

One aspect of the study of decomposition which has not been developed appropriately derives from the observation that Sylvester Theorem 6.3 can be sharpened as follows.

Theorem 6.8 *Assume $n = 1$. Assume that $T \in \mathbb{P}(Sym^d(\mathbb{C}^2))$ has a minimal decomposition A with $\ell(A) < d + 1$. Then for any other minimal decomposition B of T one has $\ell(A) + \ell(B) \geq d + 2$.*

Proof Assume on the contrary that T has a second decomposition B with $\ell(B) + \ell(A) \leq d+1$, and take the union $Z = A \cup B$. Then $\ell(Z) \leq d+1$. Then we conclude as in the proof of Theorem 6.4.

Remark 6.14 One can prove a statement similar to Theorem 6.5 under the assumption that $\langle A \rangle = \mathbb{P}^n$. Namely in this case for any other minimal decomposition B of T one has $\ell(A) + \ell(B) \geq d + n$. Details are left to the reader.

6.4   Kruskal's Criterion and Terracini's Criterion

The most famous and most used criterion for detecting the identifiability of a given tensor was proved by Kruskal in 1977 (see [21]). Kruskal's criterion was originally proved for 3way, non necessarily symmetric, tensors. The application to symmetric tensors of any size is described e.g. in [14]. We recall the result here, rephrased in terms of the geometric language.

Theorem 6.9 (Reshaped Kruskal's Criterion)   *Let* $T \in Sym^d(\mathbb{C}^{n+1})$ *and let* $A \subset \mathbb{P}^n$ *be a minimal decomposition of* T. *Fix a partition* $d = a + b + c$, *with* $0 < a \leq b \leq c$. *Write* k_a, k_b, k_c *for the Kruskal ranks of* $v_a(A)$, $v_b(A)$, $v_c(A)$ *respectively. If*

$$\ell(A) \leq \frac{k_a + k_b + k_c - 2}{2}$$

then T *has rank* $\ell(A)$ *and it is identifiable.*

Of course the efficiency of the previous criterion depends on the choice of the partition. One should observe that computing the Kruskal ranks can be demanding, for large values of d, unless the coordinates matrices of $v_a(A)$, $v_b(A)$, $v_c(A)$ have full rank. For that reason, and also for widening the range in which Kruskal's criterion applies, it is usually convenient to us a maximally unbalanced partition

Example 6.3 Consider the case $d = 4$. The unique partition is $a = b = 1, c = 2$.
  If $2 \leq \ell(A) \leq n + 1$, in the most favorable case in which $k_a = k_b = k_c = \ell(A)$, then the condition $\ell(A) \leq (k_a + k_b + k_c - 2)/2$ is automatically satisfied and Kruskal's criterion applies.
  If $n + 1 < \ell(A) \leq \binom{n+2}{2}$, then the most favorable case is $k_a = k_b = n + 1$ and $k_c = \ell(A)$. In this situation $\ell(A) \leq (k_a + k_b + k_c - 2)/2$ is equivalent to $\ell(A) \leq 2n$.
  So, one cannot hope to apply directly Kruskal's criterion, for $d = 4$, as soon as $\ell(A) > 2n$.

A direct improvement of Kruskal's criterion is impossible, unless one adds some extra test on the tensor T. Namely Kruskal's criterion (even in its reshaped version) is known to be sharp, in its maximal range.

Theorem 6.10   *For any* n, d *there exist* a, b, c *and a tensor* $T \in Sym^d(\mathbb{C}^{n+1})$ *with a minimal decomposition* A *such that the Kruskal's ranks* k_a, k_b, k_c *are maximal*

(i.e. $k_a = \min\{\ell(A), \binom{n+a}{a}\}$, and a similar equality holds for b and c), with

$$\ell(A) = \frac{k_a + k_b + k_c}{2}$$

and such that T is not identifiable.

Proof The proof is essentially due to Derksen ([16]), who proved the result in the non symmetric case. Remark 1.1 of [2] contains the observation that, when T is symmetric, then Derksen's construction provides several *symmetric* decompositions of T.

Thus, given a decomposition A of a fixed symmetric tensor T, one can test the identifiability (and the rank) of T by computing the Kruskal ranks k_a of the images of A in suitable Veronese embeddings, hoping to obtain $k_a + k_b + k_c \geq 2\ell(A) + 2$. If the inequality holds, Kruskal's theorem guarantees the identifiability of T.

Typically, the reshaped Kruskal's criterion works for small values of $\ell(A)$. To study the identifiability of tensors in a wider range, one needs to add some new test for T.

An example of a test that, together with Kruskal's test, can provide an affirmative answer for the identifiability of T, is provided by an observation which comes out from the Terracini's description of the tangent space to the set of tensors of fixed rank.

In the space $\mathbb{P}(Sym^d(\mathbb{C}^{n+1}))$, call Σ_r the set of tensors of rank r.

For small values of r, i.e. for $r(n+1) \leq \binom{n+d}{d}$, Σ_r is locally closed in the Zariski topology, i.e. it is an open subset of a projective subvariety (the r-th *secant variety* of the Veronese image $v_d(\mathbb{P}^n)$).

Consider the symmetric product $(\mathbb{P}^n)^{(r)}$. In the product

$$\mathbb{P}(Sym^d(\mathbb{C}^{n+1})) \times (\mathbb{P}^n)^{(r)}$$

consider the subvariety $A\Sigma_r$ of pairs $(T, [\{P_1, \ldots, P_r\}])$ such that the set $A = \{P_1, \ldots, P_r\}$ is mapped by v_d to a finite set which spans a subspace of dimension $r - 1$ in $\mathbb{P}(Sym^d(\mathbb{C}^{n+1}))$ (i.e. $v_d(A)$ is linearly independent) and T belongs to the span of $v_d(A)$.

The set $A\Sigma_r$, which is a quasi-projective variety, is called the *abstract secant variety* of $v_d(\mathbb{P}^n)$. The projection to the first factor maps $A\Sigma_r$ surjectively to Σ_r.

Definition 6.11 Define the r-th *secant map* s_r as the map projection to the first factor

$$s_r : A\Sigma_r \to \mathbb{P}(Sym^d(\mathbb{C}^{n+1})).$$

The image of the secant map is Σ_r. The inverse image of a tensor T of rank r in the secant map is the set of decompositions of T.

Since $v_d(\mathbb{P}^n)$ is a smooth variety, then $(\mathbb{P}^n)^{(r)}$ is smooth, outside the diagonals. Thus also $A\Sigma_r$, which is a $\mathbb{P}^{r-1}$ bundle over a subset of $(\mathbb{P}^n)^{(r)}$ which does not meet the diagonals, is smooth.

Definition 6.12 The tangent space to $A\Sigma_r$ at a point $(T, [\{P_1, \ldots, P_r\}])$ maps, in the differential of s_r, to the space $\mathscr{T}$ in $\mathbb{P}^N = \mathbb{P}(Sym^d(\mathbb{C}^{n+1}))$ spanned by the tangent spaces to $v_d(\mathbb{P}^n)$ at the points $v_d(P_1), \ldots, v_d(P_n)$. We call this space the *Terracini space* of the decomposition $A = \{P_1, \ldots, P_r\}$ of T.

The name of *Terracini space* comes from the celebrated Terracini's Lemma, which says that, for a general choice of $T \in \Sigma_r$ and for $r \leq N$, the Terracini space is the tangent space to Σ_r at T. Thus, a computation of the dimension of the Terracini space at a general point corresponds to compute the dimension of the set Σ_r of tensors of rank $r \leq N$.

Remark 6.15 The dimension of the Terracini space $\mathscr{T}$ is naturally bounded:

$$\dim(\mathscr{T}) \leq (n+1)r - 1,$$

and the equality means that the tangent spaces to $v_d(\mathbb{P}^n)$ at the points $v_d(P_i)$'s are linearly independent.

Since $A\Sigma_r$ is a $\mathbb{P}^{r-1}$ bundle over a quasi-projective variety of dimension nr, then $(n+1)r - 1 = \dim(A\Sigma_r)$. It follows that the dimension of the Terracini space equals $(n+1)r - 1$ when the differential of s_r has maximal rank.

Remark 6.16 The decomposition A of $T \in \mathbb{P}(Sym^d(\mathbb{C}^{n+1}))$ corresponds to the datum of r linear forms $L_1, \ldots, L_r$ in the polynomial ring $R = \mathbb{C}[x_0, \ldots, x_n]$.

The Terracini space can be naturally identified with the degree d homogeneous piece of the ideal in R spanned by

$$L_1^{d-1}m + \cdots + L_r^{d-1}m,$$

where m is the ideal generated by the variables.

It follows that the computation of the dimension of the Terracini space at a decomposition of T is a straightforward application of simple algorithm of *linear algebra*.

We refer to the book [20] for the (elementary) proof of this statement.

The use of the Terracini space in the computation of the identifiability of a form T is meaningful in the following situation.

Proposition 6.7 *Let A be a decomposition of T of length r and assume that there exists a non trivial family A_t of decompositions of T, such that $A_0 = A$. Then the Terracini space of A has dimension strictly smaller than $(n+1)r - 1$.*

Proof A_t determines a positive dimensional subvariety W in the fiber of s_r over T. Thus, there exists a tangent vector to $A\Sigma_r$ at $(T, [A])$, where $[A]$ is the point of the symmetric product corresponding to A, which is killed by the differential of s_r at $(T, [A])$. Then use Remark 6.15.

Now, we can introduce our strategy in finding criteria for the identifiability of symmetric tensors, which works in a range slightly wider than the Kruskal's one.

If we can prove that tensors T which are non identifiable must have a positive dimensional family of different decompositions, containing the given decomposition A, then we can check the identifiability of T by computing the dimension of the Terracini space.

The fact that non identifiable tensors have indeed a positive dimensional family of different decompositions, is false in general. It turns out, however, that this fact holds in some cases, especially when we are outside the Kruskal's numerical range, but very close to it.

A way to produce positive dimensional family of different decompositions is explained in the following:

Proposition 6.8 *Assume that a decomposition A of length r of T is contained in a projective curve $C \subset \mathbb{P}^n$ which is mapped by v_d to a space $\mathbb{P}^m$, with $m < 2r - 1$. Then there exists positive dimensional family of different decompositions A_t of T, such that $A_0 = A$.*

Proof T belongs to the span of $v_d(A)$, which is contained in the span of $v_d(C)$, which is contained in $\mathbb{P}^m$. The condition $m < 2r + 1$ implies that there is a positive dimensional family of subsets $A_t \subset C$ such that $T \in \langle v_d(A_t) \rangle$. Namely, the abstract r secant variety $A\Sigma_r^C$ of C has dimension $2r - 1$, thus all the components of the fibers of the map $A\Sigma_r^C \to \mathbb{P}^m$ are positive dimensional. $\square$

Now we can mix together the analysis of the Hilbert function, the Cayley-Bacharach condition and the computation of the dimension of the Terracini space, to produce a criterion for the identifiability of T.

Theorem 6.11 (See [2]) *Let T be a quartic form in $n + 1$ variables, and consider a decomposition A of T of length $2n + 1$.*

Assume that:

a) *the Kruskal rank of A is $n + 1$;*
b) *the Terracini space at A has (the maximal) dimension $(2n + 1)(n + 1) - 1$.*

Then T has rank $2n + 1$ and it is identifiable.

Notice that conditions a) and b) are expected to hold for a general quartic, i.e. outside a proper Zariski closed subset (of measure 0) in the space of quartics. Thus the previous theorem provides a criterion to prove the identifiability of T, except for very special tensors.

Proof We give a sketch of the proof.

First notice that, by Proposition 6.5, the set $v_2(A)$ is linearly independent, i.e. it has Kruskal rank $2n + 1$.

Call B a different decomposition of length $\leq 2n + 1$ for T, which we want to exclude. Call Z the union $Z = A \cup B$ and consider the Hilbert function of Z.

First step is to prove that Z has the Cayley-Bacharach property $CB(4)$. This is almost clear when $A \cap B = \emptyset$, while if $A \cap B \neq \emptyset$ the claim follows from Kruskal's theorem.

Next, since Z has the property $CB(4)$, by Theorem 6.1 it follows soon that $Dh_Z(3) + Dh_Z(4) + Dh_Z(5) \geq h_A(2) = 2n + 1$, so that $h_Z(2) = h_A(2) = 2n + 1$. Then one invokes the following extension of the classical Castelnuovo's Lemma:

Lemma 6.5 (See [2], Lemma 5.4)   *Let Z be a set of $r \geq 2n + 3$ points in $\mathbb{P}^n$ which impose at most $2n + 1$ conditions to quadrics. Assume that Z has a subset Z' of $2n + 1$ points in LGP. Then the entire Z is in LGP and it is contained in an irreducible rational normal curve.*

The classical formulation of Castelnuovo's lemma required that the whole set Z is in LGP, which we cannot assume in our setting, because we only know the position of A, which contains $2n + 1$ points of Z, while we have no control of the points of B. Fortunately, the extension matches exactly our requirements. Now we can turn back to the proof of the Theorem.

Since Z has a subset, namely A, which is in LGP, then it follows that Z, hence also A, sits in a rational normal curve C of $\mathbb{P}^n$. The image of C in the Veronese map v_4 spans a $\mathbb{P}^{4n}$. Hence the claim follows by Proposition 6.8.

6.5   A New Result on the Decomposition of Tensors

In this section we improve slightly Theorem 6.11, by removing the assumption that the Kruskal rank of A is $n + 1$, and replacing it by a numerical assumption on $\ell(A)$. At a certain point of the proof we will need the cohomological properties of the residue of a finite set with respect to a hyperplane. This is the unique passage in which some sophisticated algebraic machinery enters into the proof.

Let Z be a finite set in $\mathbb{P}^n$, and let H be a hyperplane. Call Z_1 the intersection $Z_1 = Z \cap H$ and call Z_2 the set:

$$Z_2 = Z \setminus Z_1 = Z \setminus (Z \cap H).$$

For obvious reasons, Z_2 is called the *residue* of Z with respect to H.

If I_Z, I_{Z_2} denote the homogeneous ideals of Z, Z_2 respectively, the multiplication by an equation of H determines an exact sequence of graded modules:

$$0 \to I_{Z_2}(1) \to I_Z(2) \xrightarrow{\rho} I_{Z_1,H}(2) \tag{6.2}$$

in which the rightmost ideal $I_{Z_1,H}$ is the homogeneous ideal of Z_1 in H.

The following result is a straightforward application of the cohomology of maps of sheaves:

Lemma 6.6 *Assume that Z_2 is linearly independent. Then the rightmost map ρ in sequence (6.2) is surjective.*

Proof The cokernel of ρ is contained in the cohomology group $H^1(\mathscr{I}_{Z_2}(1))$, where $\mathscr{I}_{Z_2}$ is the ideal sheaf of Z_2. Moreover $H^1(\mathscr{I}_{Z_2}(1))$ vanishes if Z_2 is linearly independent, because in this case the evaluation map $ev(1)$ on Z_2 determines a surjective map $\mathbb{C}^{n+1} \to \mathbb{C}^{\ell(Z_2)}$.

Remark 6.17 With the same trick, one can prove the following general statement:

Assume that the residue Z_2 of a finite set Z, with respect to a hyperplane H, is separated by forms of degree $d - 1$. Then any form of degree d in H which contains $Z_1 = Z \setminus Z_2$ can be lifted to a form of degree d in $\mathbb{P}^n$ which contains Z.

As a consequence, in the hypothesis of Lemma 6.6, it turns out that every quadric of the hyperplane H that contains Z_2 can be lifted to a quadric of $\mathbb{P}^n$ that contains Z.

We will need the following, well known remark for linearly independent sets W in a projective space $\mathbb{P}^n$:

Lemma 6.7 *Let W be a linearly independent finite set in $\mathbb{P}^n$. Then for any $Q \notin W$, there exists a quadric of $\mathbb{P}^n$ containing W and missing Q. In other words, the ideal of W is generated by quadrics.*

Proof The proof is an easy argument of linear algebra. After shrinking n we may always assume $W = \{P_1, \ldots, P_{n+1}\}$. If Q does not belong to the span of any proper subset of W, just by taking two hyperplanes containing two proper subsets we get the claim. Thus, reorder the points of W so that $P_1, \ldots, P_s$ ($s \geq 2$, $s \leq n$) is a minimal subset whose span L contains Q. Since the points are linearly independent, the span M of $P_1, \ldots, P_{s-1}, P_{s+1}$ intersects L in the span of $P_1, \ldots, P_{s-1}$, hence by minimality it does not contain Q. Similarly, the span M' of $P_s, P_{s+2}, \ldots, P_{n+1}$ intersects L only in P_n. The union of a general hyperplane containing M and a general hyperplane containing M' provides a quadric containing W and missing Q.

Now we are ready to state and proof our result.

Theorem 6.12 *Let T be a quartic form in $n + 1$ variables, and consider a decomposition A of T. Call k the Kruskal rank of A and assume that $\ell(A) \leq 2k - 1$. Assume that the Terracini space at A has (maximal) dimension $(2k - 1)(n + 1) - 1$. Then T has rank $2k - 1$ and it is identifiable.*

Notice that since $k \leq n + 1$, it follows $\ell(A) \leq 2n + 1$. Moreover, by Proposition 6.5, we know that A is separated by quadrics, i.e. $v_2(A)$ is linearly independent. This implies immediately that also $v_4(A)$ is linearly independent.

Notice also that if $\ell(A) < 2k - 1$, then A satisfies the hypothesis of the reshaped Kruskal's criterion, because in this case

$$\ell(A) \leq \frac{k + k + \ell(A) - 2}{2},$$

so that the identifiability of A follows immediately.

Thus the Theorem produces a new criterion only for $\ell(A) = 2k - 1$. Hence we assume, in the proof, that $\ell(A) = 2k - 1$.

Proof As in the proof of Theorem 6.11, we will prove that, under the assumptions, if another decomposition B of cardinality $\ell(B) \leq 2k - 1$ exists, then there exists a curve C containing A and such that $v_4(C)$ spans a space of dimension $\leq 4k - 4$, which contradicts the assumption 2).

Of course, we may assume that A spans $\mathbb{P}^n$, otherwise we simply decrease n. It follows that $2k - 1 > n$ and the difference of the Hilbert function of A is:

$$Dh_A(0) = 1, \quad Dh_A(1) = n, \quad Dh_A(2) = 2k - 2 - n.$$

Assume that a second decomposition B exists. The first step is to prove that $Z = A \cup B$ satisfies the Cayley Bacharach property $CB(4)$, which holds by following verbatim the proof of the similar statement in Theorem 6.2 of [2].

It follows then, by Theorem 6.1, that the difference of the Hilbert function of Z satisfies $Dh_Z(3) + Dh_Z(4) + Dh_Z(5) = 2k - 1$, so that in particular $\ell(B) = 2k - 1$, A, B are disjoint and the difference of the Hilbert function of Z satisfies:

$$Dh_Z(0) = 1, \quad Dh_Z(1) = n, \quad Dh_Z(2) = 2k - 2 - n.$$

Thus, summing up, one gets $h_Z(2) = h_A(2)$, i.e. all the quadrics that contain A must contain Z.

The assumption that k is the Kruskal rank of A means that any subset of k points in A is linearly independent, while there exists a subset of $k + 1$ points which generates a subspace $\Lambda = \mathbb{P}^{k-1}$. After rearranging the points, we may assume that $P_1, \ldots, P_{k+1}$ generate Λ, $P_{k+1}, \ldots, P_{k+q}$ are also contained in Λ, and $P_{k+q+1}, \ldots, P_{2k-1}$ are outside Λ. Notice that we may always assume that A is non degenerate, thus $k + q < 2k - 1$. Call Λ' the space generated by $P_{k+q+1}, \ldots, P_{2k-1}$. Any pair of hyperplanes H, H' which contain Λ, Λ' respectively, determine a quadric which contains A. It follows that all the points of B are contained either in Λ or in Λ'.

Let Q be a point of B which lies in Λ. For any subset W of $k - 1$ points among $P_1, \ldots, P_{k+q}$ consider the hyperplane L_W of Λ spanned by W. If Q belongs to no hyperplanes L_W, then there are quadrics in Λ which contain $P_1, \ldots, P_{k+q}$. Thus if H is a general hyperplane containing Λ then there are quadrics in H which contain $P_1, \ldots, P_{k+q}$ and miss Q. Since the set $P_{k+q+1}, \ldots, P_{2k-1}$ is linearly independent, by our assumption on the Kruskal rank of A, then by Lemma 6.6 one finds a quadric of $\mathbb{P}^n$ which contains A and misses Q, contradicting $h_Z(2) = h_A(2)$.

Hence, there exists a set W of $k-1$ points among $P_1, \ldots, P_{k+q}$ which spans a hyperplane L_W of Λ containing Q. Since W is linearly independent, by Lemma 6.7 one can find a quadric K in L_W that contains W and misses Q. Since, by our assumption on the Kruskal rank of A, also $\{P_1, \ldots, P_{k+q}\} \setminus W$, which contains at most k points, is linearly independent, then by Lemma 6.6 we can lift K to a quadric K' of Λ which misses Q and contains $P_1, \ldots, P_{k+q}$. As above, K' lifts to a quadric K'' which contains A and misses Q. Thus we have a contradiction with $h_Z(2) = h_A(2)$.

It follows that all the points of B belong to Λ'. In particular, the form T does not involve all the variables. After choosing carefully the coordinates $x_0, \ldots, x_n$ in $\mathbb{P}^n$, we may assume that T does not involve x_n. But then, by replacing x_n with $t x_n$ in the points of A (actually in the points of $A \cap \Lambda$), as t varies we get a family of decompositions of T which coincides with A for $t = 1$. By Proposition 6.7, this contradicts the assumption that the Terracini space has maximal dimension.

Remark 6.18 As in Section 6 of [2], one can create an algorithm that uses Theorem 6.12 to detect the identifiability of quartics of low rank. Given a symmetric decomposition of length r of a quartic

$$T = \sum_{i=1}^{r} v_4(P_i),$$

in the form of the collection of points $A = \{P_i = [\mathbf{m}_i]\}_{i=1}^{r} \subset \mathbb{P}^n$, we can apply the following algorithm for verifying that the given decomposition of T is identifiable:

1) *Kruskal's test*: compute the Kruskal rank k of A;

   S1. If $r > 2k - 1$, the criterion cannot be applied.
   S2. If $r < 2k - 1$, use the reshaped Kruskal criterion from [13, Section 6.2].
   S3. If $r = 2k - 1$, perform the:

2) *Terracini's test*: check that the dimension of $\langle T_{\mathbf{m}_1} v_4(\mathbb{C}^{n+1}), \ldots, T_{\mathbf{m}_r} v_4(\mathbb{C}^{n+1}) \rangle$ is $(2k-1)(n+1) - 1$.

If all these tests are successful, then T is of rank r and is identifiable.

Notice that the computation of the Kruskal rank of A turns out to be the heaviest step of the algorithm.

6.6   Final Remarks and Open Problems

1. We believe that the range in which the non-identifiability of tensors implies the existence of a positive dimensional family of decompositions (which can be detected by the computation of the Terracini space) goes beyond the numerical bounds given in Theorems 6.11 and 6.12.

In order to extend the previous results, however, one needs extensions of the basic Castelnuovo's Lemma 6.5. What we would need is to replace the existence of a rational normal curve, predicted by Lemma 6.5 for sets of points with special Hilbert functions, with the existence of other types of curves (elliptic, or even of higher genera), when the number of points increases.

Similar results are known in some cases (see e.g. [19, 25]), but not in a form that can be immediately applied to our situation.

We would like to stimulate further researches on the geometry of sets of points with special Hilbert functions, with the final target of an application to tensor analysis.

2. The geometric methods known so far for the study of the identifiability of specific tensors, as the Kruskal's criterion and the extension given in the previous sections, are based on the study of the geometry of a given decomposition. The idea has a basic bug: once the identifiability follows from geometric properties of a given decomposition A, then it must hold for all the tensors which lie in the span of $v_d(A)$ (at least those for which A is minimal), regardless of the coefficients that are used to produce the form T. Of course, we can expect that a similar uniform behavior holds only for small values of the rank r. When r increases, then it is natural to expect that the space $\langle v_d(A)\rangle$ contains both identifiable and non-identifiable points.

As a consequence, we need criteria for identifiability which are able to distinguish between different points of the span $\langle v_d(A)\rangle$ of a given decomposition A.

We believe that a geometric analysis of A and of its linked sets of points can produce geometric criteria which reach much further than the range of application of Kruskal's criterion.

3. A different approach to the study of the identifiability of tensors is contained in the paper [22]. The authors prove that when the space spanned by partial derivatives of the form T (the *catalecticant space*, in the terminology of [20]) meets the corresponding variety in a finite set of the expected length r, then r is the rank of T and the tensor is identifiable.

The method of partial derivatives has the advantage that it does not need to start with a given decomposition. On the other hand, for special tensors, it does not describe the geometric situation which yields the non-uniqueness of the decomposition. Furthermore, the method relies on the computation of an intersection of algebraic varieties, i.e. on methods of computer algebra, which usually cost a lot in terms of computational complexity.

We believe that a mix of the two methods, which will be the target of a forthcoming paper, will produce new, interesting developments in the theory.

4. We wonder if the analysis of tensor decomposition by means of geometric methods, related with the study of finite sets in projective spaces, can be extended beyond the case of symmetric tensors. For general tensors, the natural substitute for the Hilbert function is the multigraded Hilbert function. Indeed, for general tensors, one has only to consider the first piece of the multigraded Hilbert function, i.e. the piece bounded by the origin and the multidegree $(1, \ldots, 1)$. For

this piece of the Hilbert function, which is basically the *Segre function*, in the terminology of [12] and [6], very few is known. For instance, we do not know an analogue of Lemma 6.1, which lists the most elementary properties. A study of the Segre function, aimed to an application to tensor analysis, will probably yield several new, valuable results on the theory.

References

1. J. Alexander, A. Hirschowitz, Polynomial interpolation in several variables. J. Algebraic Geom. **4**, 201–222 (1995)
2. E. Angelini, L. Chiantini, Vannieuwenhoven N, Identifiability beyond Kruskal's bound for symmetric tensors of degree 4. Rend. Lincei Mat. Applic. **29**, 465–485 (2018)
3. E. Ballico, On the weak non-defectivity of Veronese embeddings of projective spaces. Cent. Eur. J. Math. **3**, 183–187 (2005)
4. E. Ballico, A. Bernardi, Decomposition of homogeneous polynomials with low rank. Math. Zeit. **271**, 1141–1149 (2012)
5. E. Ballico, L. Chiantini, A criterion for detecting the identifiability of symmetric tensors of size three. Differ. Geom. Appl. **30**, 233–237 (2012)
6. E. Ballico, A. Bernardi, L. Chiantini, E. Guardo, Bounds on the tensor rank. Ann. Mat. Pura Appl. **197**, 1771–1785 (2018)
7. A. Bernardi, A. Gimigliano, M. Idà, Computing symmetric rank for symmetric tensors. J. Symb. Comput. **46**, 34–53 (2011)
8. A.M. Bigatti, A.V. Geramita, J. Migliore, Geometric consequences of extremal behavior in a theorem of Macaulay. Trans. Amer. Math. Soc. **346**, 203–235 (1994)
9. J. Buczyński, A. Ginensky, J.M. Landsberg, Determinantal equations for secant varieties and the Eisenbud-Koh-Stillman conjecture. J. Lond. Math. Soc. **88**, 1–24 (2013)
10. E. Carlini, M.V. Catalisano, L. Chiantini, Progress on the symmetric Strassen conjecture. J. Pure Appl. Algebra **219**, 3149–3157 (2015)
11. L. Chiantini, C. Ciliberto, On the concept of k-secant order of a variety. J. Lond. Math. Soc. **73**, 436–454 (2006)
12. L. Chiantini, D. Sacchi, Segre functions in multiprojective spaces and tensor analysis, in *From Classical to Modern Algebraic Geometry*, ed. by G. Casnati, A. Conte, L. Gatto, L. Giacardi, M. Marchisio, A. Verra (Birkhäuser, Cham, 2016), pp. 361–374
13. L. Chiantini, G. Ottaviani, N. Vannieuwenhoven, On generic identifiability of symmetric tensors of subgeneric rank. Trans. Amer. Math. Soc. **369**, 4021–4042 (2017)
14. L. Chiantini, G. Ottaviani, N. Vannieuwenhoven, Effective criteria for specific identifiability of tensors and forms. SIAM J. Matrix Anal. Appl. **38**, 656–681 (2017)
15. E. Davis, Complete intersections of codimension 2 in $\mathbb{P}^r$ the Bezout-Jacobi-Segre theorem revisited. Rend. Sem. Mat. Univ. Politec. Torino **43**, 333–353 (1985)
16. H. Dersken, Kruskal's uniqueness inequality is sharp. Linear Algebra Appl. **438**, 708–712 (2013)
17. F. Galuppi, M. Mella, Identifiability of homogeneous polynomials and Cremona transformations. Preprint, arXiv:1606.06895 (2016)
18. A.V. Geramita, M. Kreuzer, L. Robbiano, Cayley-Bacharach schemes and their canonical modules. Trans. Amer. Math. Soc. **339**, 443–452 (1993)
19. L. Ghezzi, A generalization of the strong Castelnuovo lemma. J. Algebra **323**, 1018–1035 (2010)
20. A. Iarrobino, V. Kanev, *Power Sums, Gorenstein Algebras, and Determinantal Loci*. Springer Lecture Notes in Mathematics, vol. 1721 (Springer, Berlin, 1999)

21. J.B. Kruskal, Three-way arrays: rank and uniqueness of trilinear decompositions, with application to arithmetic complexity and statistics. Linear Algebra Appl. **18**, 95–138 (1977)
22. A. Massarenti, M. Mella, G. Staglianó, Effective identifiability criteria for tensors and polynomials. J. Symb. Comput. **87**, 227–237 (2018)
23. M. Mella, Singularities of linear systems and the Waring problem. Trans. Amer. Math. Soc. **358**, 5523–5538 (2006)
24. J. Migliore, *Introduction to Liaison Theory and Deficiency Modules*. Progress in Mathematics, vol. 165 (Birkhäuser, Basel, 1998)
25. I. Petrakiev, A step in Castelnuovo theory via Grʻo bner bases. J. Reine Angew. Math. **619**, 49–73 (2008)
26. W. Rao, D. Li, J.Q. Zhang, A tensor-based approach to L-shaped arrays processing with enhanced degrees of freedom. IEEE Signal Process Lett. **25**, 1–5 (2018)
27. Y. Shitov, A counterexample to Strassen's direct sum conjecture. Preprint, arXiv:1712.08660 (2017)
28. J.J. Sylvester, Sur une extension d'un théoréme de Clebsch relatif aux courbes du quatriéme degré. C. R. Math. Acad. Sci. Paris **102**, 1532–1534 (1886)

Chapter 7
Differential Geometry of Quantum States, Observables and Evolution

F. M. Ciaglia, A. Ibort, and G. Marmo

Abstract The geometrical description of Quantum Mechanics is reviewed and proposed as an alternative picture to the standard ones. The basic notions of observables, states, evolution and composition of systems are analysed from this perspective, the relevant geometrical structures and their associated algebraic properties are highlighted, and the Qubit example is thoroughly discussed.

7.1 Introduction

Finding a unified formalism for both Quantum Mechanics and General Relativity is an outstanding problem facing theoretical physicists. From the mathematical point of view, the structural aspects of the two theories could not be more different.

Quantum Mechanics is prevalently an algebraic theory; the transformation group, in the sense of Klein's programme, is a group of linear transformations (the group of unitary transformations on a Hilbert space for instance). General Relativity, on the other hand, sees the triumph of Differential Geometry. The covariance group of the theory is the full diffeomorphisms group of space-time.

The usual approach of non-commutative geometry consists on the algebraization of the geometrical background [9]; here, we will discuss an opposite attempt: to geometrise the algebraic description of Quantum Mechanics. In different terms, we attempt at a description of Quantum Mechanics where non-linear transformations

F. M. Ciaglia · G. Marmo
Sezione INFN di Napoli and Dipartimento di Fisica E. Pancini dell'Università Federico II di Napoli, Complesso Universitario di Monte S. Angelo, Naples, Italy
e-mail: ciaglia@na.infn.it; marmo@na.infn.it

A. Ibort (✉)
ICMAT, Instituto de Ciencias Matemáticas (CSIC-UAM-UC3M-UCM) and Depto. de Matemáticas, Univ. Carlos III de Madrid, Leganés, Madrid, Spain
e-mail: albertoi@math.uc3m.es

© Springer Nature Switzerland AG 2019
E. Ballico et al. (eds.), *Quantum Physics and Geometry*,
Lecture Notes of the Unione Matematica Italiana 25,
https://doi.org/10.1007/978-3-030-06122-7_7

153

are possible and the full diffeomorphisms group of the carrier space becomes the covariance group of the theory.

Thus we are going to introduce a "quantum differential manifold" as a carrier space for our description of Quantum Mechanics, that is, a standard smooth manifold (possibly infinite-dimensional) that will play the role of the carrier space of the quantum systems being studied. We shall use a simplifying assumption to avoid introducing infinite dimensional geometry which would go beyond the purposes of this presentation, which is conceptual rather than technical.

Of course, this idea is not new and it has been already explored earlier. Just to mention a few, we may quote the early attempts by Kibble [21], the essay by Asthekar and Schilling [2], the mathematical foundations laid by Cirelli et al. [8] and the systematic search for a geometric picture led by Marmo (see for instance early ideas in the subject in the book [12] and some preliminary results in [16] or the review [11]). This work is a continuation of this line of thought and contains a more comprehensive description of such attempt.

Let us briefly recall first the various pictures of Quantum Mechanics, emphasising the algebraic structures present in their description.

7.1.1 On the Many Pictures of Quantum Mechanics

As it is well known, modern Quantum Mechanics was first formulated by Heisenberg as matrix mechanics, immediately after Schrödinger formulated his wave mechanics. These pictures got a better mathematical interpretation by Dirac [10] and Jordan [4, 20] with the introduction of Hilbert spaces and Transformation Theory. Further, a sound mathematical formulation was provided by von Neumann [27].

In all of these pictures and descriptions, the principle of analogy with classical mechanics, as devised by Dirac, played a fundamental role. The canonical commutation relations (CCR) were thought to correspond or to be analogous to the Poisson Brackets on phase space. Within the rigorous formulation of von Neumann, domain problems were identified showing that at least one among position and momentum operators should be an unbounded operator [29]. To tackle these problems, Weyl introduced an "exponentiated form" of the commutation relations in terms of unitary operators [28], i.e., a projective unitary representation of the symplectic Abelian vector group, interpreted also as a phase-space with a Poisson Bracket. The C^*-algebra of observables, a generalization of the algebraic structure emerging from Heisenberg picture, would be obtained as the group-algebra of the Weyl operators.

7.1.2 Dirac-Schrödinger vs. Heisenberg-Weyl Picture

Even if commonly used, there is not an universal interpretation of the term "picture" used above as applied to a particular mathematical embodiment of the axioms used in describing quantum mechanical systems. The description of any physical system requires the identification of:

1. States.
2. Observables.
3. A probability interpretation.
4. Evolution.
5. Composition of systems.

Thus, in this work, a "picture" for a quantum mechanical system will consist of a mathematical description of: (1) a collection of states $\mathscr{S}$; (2) a collection of measurable physical quantities or observables $\mathscr{A}$; (3) a statistical interpretation of the theory, that is, a pairing:

$$\mu \colon \mathscr{S} \times \mathscr{A} \to \mathbf{Bo}(\mathbb{R}) \tag{7.1}$$

where $\mathbf{Bo}(\mathbb{R})$ is the set of Borel probability measures on the real line and if $\rho \in \mathscr{S}$ denotes a state of the system and a an observable, then, the pairing $\mu(\rho, a)(\Delta)$ is interpreted as the probability $P(\Delta|a, \rho)$ that the outcome of measuring the observable a lies in the Borelian set $\Delta \subset \mathbb{R}$ if the system is in the state ρ. In addition to these "kinematical" framework a "picture" of a quantum system should provide (4) a mathematical description of its dynamical behaviour and (5) prescription for the composition of two different systems.

7.1.2.1 Dirac-Schrödinger Picture

Thus, for instance, in the Dirac-Schrödinger picture with any physical system we associate a complex separable Hilbert space $\mathscr{H}$. The (pure) states of the theory are given by rays in the Hilbert space, or equivalently by rank-one orthogonal projectors $\rho = |\psi\rangle\langle\psi|/\langle\psi|\psi\rangle$ with $|\psi\rangle \in \mathscr{H}$. Observables are Hermitian or self-adjoint operators a (bounded or not) on the Hilbert space and the statistical interpretation of the theory is provided by the resolution of the identity E (or spectral measure $E(d\lambda)$) associated to the observable by means of the spectral theorem, $a = \int \lambda E(d\lambda)$. Thus the probability $P(\Delta|a, \rho)$ that the outcome of measuring the observable A when the system is in the state ρ would lie in the Borel set $\Delta \subset \mathbb{R}$, is given by:

$$P(\Delta|a, \rho) = \int_\Delta \mathrm{Tr}(\rho E(d\lambda)) . \tag{7.2}$$

Moreover the evolution of the system is dictated by a Hamiltonian operator H by means of Schrödinger's equation:

$$i\hbar \frac{d}{dt}|\psi\rangle = H|\psi\rangle .$$

Finally, if $\mathscr{H}_A$ and $\mathscr{H}_B$ denote the Hilbert spaces corresponding to two different systems, the composition of them has associated the Hilbert space $\mathscr{H}_A \otimes \mathscr{H}_B$.

7.1.2.2 Heisenberg-Born-Jordan

In contrast, in the Heisenberg-Born-Jordan picture a unital C^*-algebra $\mathscr{A}$ is associated to any physical system. Observables are real elements $a = a^*$ in $\mathscr{A}$ and states are normalised positive linear functionals ρ on $\mathscr{A}$:

$$\rho(a^*a) \geq 0, \quad \rho(1_\mathscr{A}) = 1,$$

where $1_\mathscr{A}$ denotes the unit of the algebra $\mathscr{A}$. The GNS construction of a Hilbert space $\mathscr{H}_\rho$, once a state ρ is chosen, reproduces the Dirac-Schrödinger picture. Similar statements can be made with respect to the statistical interpretation of the theory. Given a state ρ and an observable $a \in \mathscr{A}$, the pairing μ between states and observables, Eq. (7.1), required to provide a statistical interpretation of the theory is provided by the spectral measure associated to the Hermitian operator $\pi_\rho(a)$ determined by the canonical representation of the C^*-algebra $\mathscr{A}$ in the Hilbert space $\mathscr{H}_\rho$ obtained by the GNS construction with the state ρ. Alternatively, given a resolution of the identity, i.e., in the discrete setting, $E_j \in \mathscr{A}$ such that $E_i \cdot E_j = \delta_{ij} E_j$, and $\sum_j E_j = 1_\mathscr{A}$, we define $p_j(\rho) = \rho(E_j) \geq 0, \sum_j p_j(\rho) = 1$. This provides the probability function of the theory.

The evolution of the theory is defined by means of a Hamiltonian $H \in \mathscr{A}$, $H = H^*$, by means of Heisenberg equation:

$$i\hbar \frac{da}{dt} = [H, a].$$

Finally, composition of two systems with C^*-algebras $\mathscr{A}_A$ and $\mathscr{A}_B$ would be provided by the tensor product C^*-algebra $\mathscr{A}_{AB} = \mathscr{A}_A \otimes \mathscr{A}_B$ (even though there is not a unique completion of the algebraic tensor product of C^*-algebras in infinite dimensions, a problem that will not concern us here as the subsequent developments are restricted to the finite-dimensional situation in order to properly use the formalism of differential geometry).

7.1.2.3 Other Pictures

The Dirac-Schrödinger and Heisenberg-Born-Jordan are far from being the only two pictures of Quantum Mechanics. Other pictures include the Weyl-Wigner picture, where an Abelian vector group V with an invariant symplectic structure ω is required to possess a projective unitary representation:

$$W: V \to \mathscr{U}(\mathscr{H}), \quad W(v_1)W(v_2)W(v_1)^\dagger W(v_2)^\dagger = e^{i\omega(v_1, v_2)} 1_\mathscr{H}.$$

The tomographic picture [18] has been developed in the past few years and uses a tomographic map U and includes the so called Wigner picture based on the use of pseudo probability distributions on phase space; a picture based on the choice of a family of coherent states has also been partially developed recently (see for instance [5]). A deep and careful reflexion would be required to analyse the 'Lagrangian picture' proposed by Dirac and Schwinger, that would be treated elsewhere.

7.2   A Geometric Picture of Quantum Mechanics

As it was discussed in the introduction, the proposal discussed in this work departs from the other ones in setting a geometrical background for the theory, so that the group of natural transformations of the theory becomes the group of diffeomorphisms of a certain carrier space. In this picture, the carrier space $\mathscr{P}$ we associate with every quantum system is the Hilbert manifold provided by the complex projective space. By taking this point of view, states and observables should be defined by means of functions on $\mathscr{P}$. This carrier space comes equipped with a Kählerian structure, i.e., a symplectic structure, a Riemannian structure and a complex structure. All three tensors, pairwise, satisfy a compatibility condition, two of them will determine the third one. We will show how to implement on this carrier space the minimalist requirements stated at the beginning of Sect. 7.1.2.

As it was commented before, to properly use the formalism of differential geometry, we shall restrict our considerations to finite dimensional complex projective spaces. We believe that, at this stage, considering infinite-dimensional systems would introduce a significant amount of technical difficulties without adding any relevant improving in the exposition of the structural aspects of the ideas we want to convey. A more thorough analysis of the infinite-dimensional case will be pursued elsewhere.

It is our hope that the "geometrization" of Quantum Mechanics can be useful to understand under which conditions any "generalized" geometrical quantum theory reduces to the conventional Dirac-Schrödinger picture.

The Carrier Space $\mathscr{P}$ is taken to be the complex projective space $\mathbb{CP}(\mathscr{H})$ associated with the n-dimensional complex Hilbert space $\mathscr{H}$. This a Hilbert manifold with a Kähler structure even in the infinite-dimensional case [8]. The Kähler structure of $\mathscr{P}$ consists of a complex structure J, a metric tensor g called the Fubini-Study metric, and a symplectic form ω. These tensor fields are mutually related according to the following compatibility condition

$$g\,(X\,,J(Y)) = \omega\,(X\,,Y)\,, \tag{7.3}$$

where X and Y are arbitrary vector fields on $\mathscr{P}$. The complex sum $h = g + \iota\omega$ is a Hermitian tensor on $\mathscr{P}$. Following [2, 11], we consider the canonical projection

$\pi : \mathcal{H}_0 \rightarrow \mathbb{CP}(\mathcal{H}) \equiv \mathscr{P}$ associating to each non-zero vector $\psi \in \mathcal{H}_0{}^1$ its ray $[\psi] \in \mathscr{P}$ and the Hermitian tensor:

$$\tilde{h} = \pi^* h = \frac{\langle d\psi | d\psi \rangle}{\langle \psi | \psi \rangle} - \frac{\langle d\psi | \psi \rangle \langle \psi | d\psi \rangle}{\langle \psi | \psi \rangle^2} . \tag{7.4}$$

The real part of this tensor is symmetric, and defines the pullback to $\mathcal{H}$ of the Fubini-Study metric g, while the imaginary part is antisymmetric and defines the pullback to $\mathcal{H}$ of the symplectic form ω.

We stress that, because our description is tensorial, we may perform any non-linear transformation without affecting the description of the theory. For instance, introducing an orthonormal basis $\{|e_j\rangle\}_{j=1,\dots,n}$ in $\mathcal{H}$, we can write every normalized vector $|\psi\rangle$ in $\mathcal{H}$ as a probability amplitude $|\psi\rangle = \sqrt{p_j}\, e^{i\varphi_j} |e_j\rangle$, $p_j \geq 0$ for all j. Clearly, $(p_1, \dots, p_n)$ is a probability vector, that is, $\sum p_j = 1$, while $e^{i\varphi_j}$ is a phase factor. Then we can compute $\tilde{h}$ in this nonlinear coordinate system obtaining:

$$\tilde{h} = \frac{1}{4} \left[\langle d(\ln \mathbf{p}) \otimes d(\ln \mathbf{p}) \rangle_{\mathbf{p}} - \langle d(\ln \mathbf{p}) \rangle_{\mathbf{p}} \otimes \langle d(\ln \mathbf{p}) \rangle_{\mathbf{p}} \right] +$$

$$+ \langle d\boldsymbol{\varphi} \otimes d\boldsymbol{\varphi} \rangle_{\mathbf{p}} - \langle d\boldsymbol{\varphi} \rangle_{\mathbf{p}} \otimes \langle d\boldsymbol{\varphi} \rangle_{\mathbf{p}} + \frac{i}{2} \left[\langle d(\ln \mathbf{p}) \wedge d\boldsymbol{\varphi} \rangle_{\mathbf{p}} - \langle d(\ln \mathbf{p}) \rangle_{\mathbf{p}} \wedge \langle d\boldsymbol{\varphi} \rangle_{\mathbf{p}} \right] , \tag{7.5}$$

where $\langle \cdot \rangle_{\mathbf{p}}$ denotes the expectation value with respect to the probability vector $\mathbf{p}$. Note that (the pullback of) g is composed of two terms, the first one is equivalent to the Fisher-Rao metric on the space of probability vectors $(p_1 \dots p_n)$, while the second term can be interpreted as a quantum contribution to the Fisher-Rao metric due to the phase of the state [13].

Given a smooth function $f \in \mathscr{F}(\mathscr{P})$, we denote by X_f, Y_f the vector fields given respectively by: $X_f = \Lambda(df)$, and $Y_f = R(df)$, where $\Lambda = \omega^{-1}$ and $R = g^{-1}$. The vector fields X_f will be called Hamiltonian vector fields and Y_f, gradient vector fields. Note that the compatibility condition among ω, g and J allows us to write $Y_f = J(X_f)$.

The special unitary group $SU(\mathcal{H})$ acts naturally on $\mathscr{P} = \mathbb{CP}(\mathcal{H})$ by means of isometries of the Kähler structure. The infinitesimal version of this action is encoded in a set of Hamiltonian vector fields $\{X_A \mid \mathbf{A} \in \mathfrak{su}(\mathcal{H})\}$ such that they close on a realization of the Lie algebra $\mathfrak{su}(\mathcal{H})$ of $SU(\mathcal{H})$. This means that, given $\mathbf{A}, \mathbf{B} \in \mathfrak{su}(\mathcal{H})$, there are Hamiltonian vector fields X_A, X_B on $\mathscr{P}$ such that $[X_A, X_B] = -X_{[A,B]}$ [1].

---

[1] $\mathcal{H}_0$ denotes the Hilbert space $\mathcal{H}$ with the zero vector removed.

The fact that $SU(\mathcal{H})$ acts preserving the Kähler structure means that the Hamiltonian vector fields for the action preserve ω, g and J, that is, $\mathscr{L}_{X_A}\omega = \mathscr{L}_{X_A}g = \mathscr{L}_{X_A}J = 0$ for every X_A. Note that this is not true for a Hamiltonian vector field X_f associated with a generic smooth function f on $\mathscr{P}$.

It is interesting to note that the Hamiltonian vector fields X_A together with the gradient vector fields $Y_A = J(X_A)$ close on a realization of the Lie algebra $\mathfrak{sl}(\mathcal{H})$, that is, the Lie algebra of the complex special linear group $SL(\mathcal{H})$ which is the complexification of $SU(\mathcal{H})$. In order to see this, we recall the definition of the Nijenhuis tensor N_J associated with the complex structure J (see definition 2.10, and equation 2.4.26 in [24]):

$$N_J(X, Y) = \left(\mathscr{L}_{J(X)}(T)\right)(Y) - (J \circ \mathscr{L}_X(J))(Y), \tag{7.6}$$

where X, Y are arbitrary vector fields on $\mathscr{P}$. A fundamental result in the theory of complex manifold is that the $(1, 1)$-tensor field J defining the complex structure of a complex manifold must have vanishing Nijenhuis tensor [25]. This means that the complex structure J on $\mathscr{P}$ is such that $N_J = 0$, which means:

$$\left(\mathscr{L}_{J(X)}(J)\right)(Y) = (J \circ \mathscr{L}_X(J))(Y), \tag{7.7}$$

where X, Y are arbitrary vector fields on $\mathscr{P}$. In particular, if we consider the Hamiltonian vector field X_A, we know that $\mathscr{L}_{X_A}J = 0$, and thus:

$$\left(\mathscr{L}_{J(X_A)}(J)\right)(Y) = 0 \tag{7.8}$$

for every vector field Y on $\mathscr{P}$. Eventually, we prove the following:

Proposition 7.1 Let $\mathbf{A}$, $\mathbf{B}$ be generic elements in the Lie algebra $\mathfrak{su}(\mathcal{H})$ of $SU(\mathcal{H})$ The Hamiltonian and gradient vector fields X_A, X_B, Y_A, Y_B on $\mathscr{P}$ close on a realization of the Lie algebra $\mathfrak{sl}(\mathcal{H})$, that is, the following commutation relations among Hamiltonian and gradient vector fields hold:

$$[X_A, X_B] = -X_{[A,B]}, \qquad [X_A, Y_B] = -Y_{[A,B]}, \qquad [Y_A, Y_B] = X_{[A,B]}. \tag{7.9}$$

Proof The first commutator follows directly from the fact that there is a left action of $SU(\mathcal{H})$ on $\mathscr{P}$ of which the Hamiltonian vector fields X_A are the fundamental vector fields. Regarding the second commutator, we recall that $Y_A = J(X_A)$ and that $\mathscr{L}_{X_A}J = 0$, so that:

$$\begin{aligned}
[X_\mathbf{A}, Y_\mathbf{B}] &= \mathscr{L}_{X_A}(J(X_\mathbf{B})) = \\
&= \left(\mathscr{L}_{X_A}J\right)(X_\mathbf{B}) + J\left(\mathscr{L}_{X_A}X_\mathbf{B}\right) = \\
&= J\left([X_\mathbf{A}, X_\mathbf{B}]\right) = -Y_{[A,B]}
\end{aligned} \tag{7.10}$$

as claimed. Finally, using Eq. (7.8) together with the fact that $J \circ J = -\mathrm{Id}$ because it is a complex structure, we obtain:

$$
\begin{aligned}
[Y_{\mathbf{A}}, Y_{\mathbf{B}}] &= \mathscr{L}_{J(X_{\mathbf{A}})}\left(J(X_{\mathbf{B}})\right) = \\
&= \left(\mathscr{L}_{J(X_{\mathbf{A}})}(J)\right) X_{\mathbf{B}} + J\left(\mathscr{L}_{J(X_{\mathbf{A}})}X_{\mathbf{B}}\right) = \\
&= J\left([Y_{\mathbf{A}}, X_{\mathbf{B}}]\right) = X_{[\mathbf{A},\mathbf{B}]}
\end{aligned}
\tag{7.11}
$$

as claimed.   □

Since $\mathscr{P}$ is a compact manifold, all vector fields are complete, in particular, the Hamiltonian and gradient vector fields of Proposition 7.1 are complete. This means that the realization of the Lie algebra $\mathfrak{sl}(\mathscr{H})$ integrates to an action of $SL(\mathscr{H})$ on $\mathscr{P}$. We will see that this action on $\mathscr{P}$ allows us to define an action of $SL(\mathscr{H})$ on the space $\mathscr{S}$ of quantum states.

Remark 7.1 Instead of the complex projective space, we may as well have started with a generic homogeneous space of $SU(\mathscr{H})$ as a carrier manifold. Every such manifold is a compact Kähler manifold, and the Hamiltonian and gradient vector fields associated with elements in $\mathfrak{su}(\mathscr{H})$ close on a realization of the Lie algebra of $SL(\mathscr{H})$ which integrates to a group action. Indeed, all we need to prove an analogue of Proposition 7.1 is a Kähler manifold on which $SU(\mathscr{H})$ acts by means of isometries of the Kähler structure.

The complex projective space may be selected requiring the holomorphic sectional curvature of $\mathscr{P}$ to be constant and positive. Indeed, from the Hawley-Isuga Theorem [17, 19], it follows that complex projective spaces are the only (connected and complete) Kähler manifolds of constant and positive holomorphic sectional curvature (in our setting equal to $2/\hbar$) up to Kähler isomorphisms.

Observables are real functions $f \in \mathscr{F}(\mathscr{P})$ satisfying:

$$
\mathscr{L}_{X_f} R = 0,
\tag{7.12}
$$

i.e., such that the Hamiltonian vector fields defined by them are isometries for the symmetric tensor $R = g^{-1}$. In particular, if F is a complex-valued function on $\mathscr{P}$ generating a complex-valued Hamiltonian vector field X_F which is Killing for g (hence for R), then there necessarily exist a, b Hermitian operators such that [2, 8, 11, 26]:

$$
F([\psi]) = \frac{\langle\psi|a|\psi\rangle}{\langle\psi|\psi\rangle} + \iota\frac{\langle\psi|b|\psi\rangle}{\langle\psi|\psi\rangle}.
\tag{7.13}
$$

This result is interesting but not unexpected, Hamiltonian vector fields are infinitesimal generators of symplectic transformations. If they also preserve the Euclidean metric, they must be infinitesimal generators of rotations, then the intersection of symplectic and rotations are unitary transformations, whose infinitesimal generators are (skew) Hermitian matrices. From what we have just seen it follows that the

observables can be identified with the expectation-value functions:

$$e_a([\psi]) = \frac{\langle \psi | a | \psi \rangle}{\langle \psi | \psi \rangle} \tag{7.14}$$

with a a Hermitian operator on $\mathscr{H}$ (notice that, consistently, $A = \imath a$ is an element in the Lie algebra $\mathfrak{su}(\mathscr{H})$ of the unitary group $SU(\mathscr{H})$). We will denote the family of observables as $\mathscr{K}(\mathscr{P})$ or simply $\mathscr{K}$ for short.

We find out that, under adequate conditions, the family of functions $\mathscr{K}$ constitutes a Lie-Jordan algebra. Indeed, the space of Kählerian functions, that is, those satisfying condition (7.12) above, because of Hawley-Igusa theorem carries a natural C^*-algebra structure and its real part a Lie-Jordan one ([3, 17, 19], [22, Thm. 7.9]). By using a GNS construction for the C^*-algebra we get a Hilbert space, returning to the Dirac-Schrödinger picture.

By using $\Lambda(\omega)$ and $R(g)$ we can define the following brackets among functions on $\mathscr{P}$:

$$\{f_1, f_2\} := \Lambda(\mathrm{d}f_2, \mathrm{d}f_1) = \omega(X_{f_1}, X_{f_2}) = X_{f_2}(f_1), \tag{7.15}$$

$$(f_1, f_2) := R(\mathrm{d}f_1, \mathrm{d}f_2) = g(Y_{f_1}, Y_{f_2}). \tag{7.16}$$

The antisymmetric bracket $\{\cdot, \cdot\}$ is a Poisson bracket since it is defined starting from a symplectic form. Furthermore, being $[X_{f_1}, X_{f_2}] = -X_{\{f_1, f_2\}}$ for every smooth functions f_1, f_2 on $\mathscr{P}$, and since $[X_A, X_B] = -X_{[A,B]}$ for the Hamiltonian vector fields associated with A, $B \in \mathfrak{su}(\mathscr{H})$, we have:

$$-X_{[A,B]} = [X_A, X_B] = [X_{f_a}, X_{f_b}] = -X_{\{f_a, f_b\}}, \tag{7.17}$$

where we have switched the notation e_a to f_a to make formulas more familiar and readable. From (7.17) it follows:

$$\{f_a, f_b\} = f_{\imath[a,b]}, \tag{7.18}$$

where we used the fact that $A = \imath a$ and $B = \imath b$. This means that $(\mathscr{K}(\mathscr{P}), \{\cdot, \cdot\})$ is a Lie algebra.

On the other hand, a direct computation [11] shows that:

$$(f_a, f_b) := R(\mathrm{d}f_a, \mathrm{d}f_b) = g(Y_a, Y_b) = f_{a \odot b} - f_a \cdot f_b, \tag{7.19}$$

where $a \odot b = ab + ba$. Then, we may define the symmetric bracket:

$$< f_1, f_2 > := (f_1, f_2) + f_1 \cdot f_2 \tag{7.20}$$

so that on the subspace of **observables** we have:

$$< f_a, f_b > = f_{a \odot b} . \tag{7.21}$$

Because of the properties of the symmetric product $\odot$ on Hermitian operators, the bracket $< \cdot, \cdot >$ turns out to be a Jordan product. Furthermore, the set of **observables** endowed with the antisymmetric product $\{\cdot, \cdot\}$ and the symmetric product $< \cdot, \cdot >$ is a Lie-Jordan algebra [6, 7, 14]. By complexification, that is, considering complex-valued functions $F_A = f_{a_1} + \iota f_{a_2}$ for some Hermitian a_1, a_2, we obtain a realization of the C^*-algebra $\mathscr{B}(\mathscr{H})$ by means of smooth functions on $\mathscr{P} = \mathbb{CP}(\mathscr{H})$ according to [8, 11]:

$$
\begin{aligned}
F_A \star F_B &:= \frac{1}{2} \left(F_A \cdot F_B + (F_A, F_B) + \iota \{F_A, F_B\} \right) = \\
&= \frac{1}{2} \left(< F_A, F_B > + \iota \{F_A, F_B\} \right) = F_{AB} .
\end{aligned}
\tag{7.22}
$$

We may extend this product to arbitrary complex-valued functions obtaining a $\star$-product.

Because we are in finite dimensions we can consider the critical points of the observables (that is, expectation value functions). An observable is said to be generic if all critical points are isolated. The values of the observable function at these critical points constitute the spectrum of the observable. The set of critical point of a generic observable may be thought of as the geometrical version adapted to $\mathscr{P} \equiv \mathbb{CP}(\mathscr{H})$ of an orthogonal resolution of the identity on $\mathscr{H}$. If a critical point is not isolated, the critical set is actually a submanifold of (real) even dimension. If the observable has value zero in some critical set, this set is a complex projective space.

We postpone a complete discussion of the critical values of a given observable and restrict our analysis to generic observables. With the help of any generic observable we can now define **quantum states**. The space $\mathscr{S}$ of **quantum states** is a subset of $\mathscr{H}$ whose elements are defined as follows. A function in $\mathscr{H}$ will define a state if its evaluation on the set of isolated critical points of any generic observable will be a probability distribution on n-elements, i.e., a discrete probability distribution. In a certain sense, we may think of quantum states (in finite dimensions) as a sort of noncommutative generalization of discrete probability distributions. Essentially, **quantum states** are identified with the expectation-value functions

$$e_\rho([\psi]) = \frac{\langle \psi | \rho | \psi \rangle}{\langle \psi | \psi \rangle}$$

associated with density operators, that is, $\rho \in \mathscr{B}(\mathscr{H})$, $\rho = \rho^\dagger$, $\langle \psi | \rho | \psi \rangle \geq 0$ for all $|\psi\rangle \in \mathscr{H}$, and $\mathrm{Tr}(\rho) = 1$. In the infinite-dimensional case ρ must be trace-class in order for this last requirement to make sense.

On the other hand, the expectation value function associated with a quantum state will define a "continuous" probability distribution on the carrier space provided by the complex projective space. Essentially, a quantum state is identified with an observable (expectation value function) $e_\rho \in \mathcal{K}$ such that $e_\rho([\psi]) \geq 0$ for all $[\psi] \in \mathcal{P}$ ($\rho \in \mathcal{B}(\mathcal{H})$ is a positive semidefinite operator), and ($\mathrm{Tr}\,\rho = 1$):

$$\int_{\mathcal{P}} e_\rho \, dv_\omega = 1 , \qquad (7.23)$$

where $dv_\omega = \omega^n$ is the symplectic volume form normalized by $\int_{\mathcal{P}} dv_\omega = 1$. This point of view would be closer to the point of view taken by Gelfand and Naimark to define states as functions of positive-type in the group algebra of any Lie group. They would be of positive-type when pulled back to the group from the homogeneous space. It is clear that they form a convex body whose extremals are the pure quantum states.

In this context, the pairing map between quantum states and observables given by:

$$E(e_\rho, f_a) = \int_{\mathcal{P}} f_a \, e_\rho \, dv_\omega \qquad (7.24)$$

is interpreted as the mean value for the outcome of a measurement of the observable f_a on the quantum state e_ρ.

Remark 7.2 In the infinite-dimensional case we must pay attention to topological and measure-theoretical issues since **quantum states** are required to be measurable with respect to the symplectic measure v_ω, while **observables** are not.

We may define the following map:

$$\mathcal{P} \ni [\psi] \mapsto \rho_\psi := \frac{|\psi\rangle\langle\psi|}{\langle\psi|\psi\rangle} \in \mathcal{B}(\mathcal{H}) . \qquad (7.25)$$

This map allows us to identify the points of $\mathcal{P}$ with rank-one projectors on $\mathcal{H}$, and, since rank-one projector are density operators, we identify the points in the carrier space $\mathcal{P}$ with particular quantum states. These quantum states are precisely the extremal points of the convex set $\mathcal{S}$ of all quantum states, that is, pure quantum states. In this context, the expectation value function e_{ρ_ψ} associated with the pure quantum state ρ_ψ encodes the transition probabilities between the normalized vector $|\tilde{\psi}\rangle$ associated with $|\psi\rangle$ and every other normalized vector $|\tilde{\phi}\rangle$ in $\mathcal{H}$:

$$e_{\rho_\psi}([\phi]) = \frac{\langle\phi|\rho_\psi|\phi\rangle}{\langle\phi|\phi\rangle} = \frac{\langle\phi|\psi\rangle\langle\psi|\phi\rangle}{\langle\phi|\phi\rangle\langle\psi|\psi\rangle} = |\langle\tilde{\psi}|\tilde{\phi}\rangle|^2 . \qquad (7.26)$$

Recalling that a quantum state is a positive function on $\mathcal{P}$, that is, $e_\rho \geq 0$, we can define the rank of a **quantum state** as the codimension of the closed submanifold $e_\rho^{-1}(0) \subset \mathcal{P}$. With this definition, it is clear that the rank is invariant under the group of diffeomorphisms. As a matter of fact it is possible to show that the complex special Lie group $SL(\mathcal{H})$ acting on $\mathcal{P}$ by means of diffeomorphisms acts transitively on the space of states with the same rank, providing in this way a stratification of the space of states. To be able to change the rank of a state, to describe decoherence, we need to use semigroups.

Writing $|\psi_G\rangle \equiv G|\psi\rangle$ with $G \in SL(\mathcal{H})$, the action of the special linear group $SL(\mathcal{H})$ on the carrier space $\mathcal{P}$ reads:

$$[G]: [\psi] \mapsto [G]([\psi]) = [\psi_G].   \tag{7.27}$$

In terms of the rank-one projector ρ_ψ we have:

$$\rho_\psi \mapsto G \cdot \rho_\psi = \frac{G|\psi\rangle\langle\psi|G^\dagger}{\langle\psi|G^\dagger G|\psi\rangle} = \frac{G\rho_\psi G^\dagger}{\mathrm{Tr}(G^\dagger \rho_\psi G)}.   \tag{7.28}$$

We may generalize this action to any density operator by setting:

$$G \cdot \rho = \frac{G^\dagger \rho G}{\mathrm{Tr}(G^\dagger \rho G)}.   \tag{7.29}$$

However, because the action is nonlinear this is an assumption that cannot derived from the action on rank-one projectors. By means of this action we would get an orbit of density operators and thus an orbit of probability distributions once we identify the density operators with their associated expectation-value functions. Each orbit being characterised by the rank of ρ. For a system with n levels (dim $\mathcal{H} = n$) we would get n different orbits. The one of maximal dimension would be the bulk, while the boundary of the closed convex body $\mathcal{S}$ of quantum states would be the union of orbits of dimensions less than n. The geometry of $\mathcal{S}$ as developed in [6, 7, 16] will be exposed in Sect. 7.2.1.

The **statistical interpretation** of the theory is provided by a geometric measure. The idea is to extend the notion of spectral measure to a geometric manifold as it was proposed for instance by Skulimowski [26] in the case of the complex projective space $\mathcal{P}$. Thus we may use a slightly extended notion defined as: a geometric positive-operator-valued measure (GPOV-measure) on a space of states of a geometric quantum theory is a map $p \colon \mathcal{B}(\mathbb{R}) \to \mathcal{K}(\mathcal{P})$ (where $\mathcal{B}(\mathbb{R})$ denotes the σ-algebra of Borelian sets in $\mathbb{R}$) such that:

1. Positivity monotonicity and normalization:

$$0 \leq p(\emptyset)([\psi]) \leq p(\Delta)([\psi]) \leq p(\mathbb{R})([\psi]) = 1.$$

2. Additivity: μ is additive, i.e.,

$$p(\cup_{k=1}^{n}\Delta_k)([\psi]) = \sum_{k=1}^{n} p(\Delta_k)([\psi])\,,$$

for all $[\psi] \in \mathscr{P}$, $n \in \mathbb{N}$, Δ_k, $k = 1, \ldots, n$, disjoint Borel sets on $\mathbb{R}$.

Thus, consider for instance a GPOV-measure p with finite support, supp $p = \{\lambda_1, \ldots, \lambda_r\}$, then the statistical interpretation of the theory will be provided, as in the standard pictures, by the probability distribution $p_k([\psi]) = p(\{\lambda_k\})([\psi]) \geq 0$, $\sum_k p_k([\psi]) = 1$.

In general a GPOV-measure p will be provided by any observable e_a by means of the corresponding spectral measure $E_a(d\lambda)$ associated to the Hermitian operator a, that is

$$p(\Delta)([\psi]) = \int_{\Delta} \mathrm{Tr}\ (\rho_\psi)E(d\lambda)) = \int_{\Delta} \frac{\langle \psi|E(d\lambda)|\psi \rangle}{\langle \psi|\psi \rangle}\,,$$

in accordance with the probabilistic interpretation of a physical theory, Eq. (7.1), and the standard pictures, Eq. (7.2).

Hamiltonian evolution, or evolution of closed systems, will be defined by the Hamiltonian vector field X_h associated with the observable h, that is:

$$\frac{df}{dt} = X_h(f)\,.$$

We call the observable h the Hamiltonian function for the evolution.

The composition of systems will be discussed in Sect. 7.3.

7.2.1 Quantum States and Open Systems

The geometry of $\mathscr{S}$ as a closed convex body in the affine ambient space $\mathfrak{T}_1$ of Hermitian operators on $\mathscr{H}$ with trace equal to 1 has been extensively developed in [6, 7]. In these works, it is shown that there exist two bivector fields Λ and $\mathscr{R}$ on $\mathfrak{T}_1$ by means of which the infinitesimal version of the action of $SL(\mathscr{H})$ on $\mathscr{S}$ may be recovered in terms of Hamiltonian and gradient-like vector fields. In this case, the Poisson bivector field Λ does not come from a symplectic structure, and the symmetric bivector field $\mathscr{R}$ is not invertible (there is no metric tensor $g = \mathscr{R}^{-1}$).

7.2.1.1 The Qubit

We will briefly recall here the results of [6, 7] concerning the geometry of the space of all states, pure or mixed for the qubit. Every 2 by 2 Hermitian matrix A may be written in the form:

$$A = \begin{bmatrix} x_0 + x_3 & x_1 - ix_2 \\ x_1 + ix_2 & x_0 - x_3 \end{bmatrix},$$

or, written as combination of Pauli matrices:

$$\sigma_0 = \begin{bmatrix} 1 & 0 \\ 0 & 1 \end{bmatrix}, \quad \sigma_1 = \begin{bmatrix} 0 & 1 \\ 1 & 0 \end{bmatrix}, \quad \sigma_2 = \begin{bmatrix} 0 & -i \\ i & 0 \end{bmatrix}, \quad \sigma_3 = \begin{bmatrix} 1 & 0 \\ 0 & -1 \end{bmatrix},$$

we get:

$$A = a_0\sigma_0 + a_1\sigma_1 + a_2\sigma_2 + a_3\sigma_3 .$$

In particular it is well known that any density operator ρ, that is Tr $\rho = 1, 0 \leq \rho^2 \leq \rho$ can be written as:

$$\rho = \frac{1}{2}(\sigma_0 + \mathbf{x} \cdot \boldsymbol{\sigma}), \quad ||\mathbf{x}|| \leq 1 .$$

Thus the space $\mathscr{S}$ of all qubit states is the Bloch's ball in $\mathbb{R}^3$:

$$\mathscr{S} = \{\mathbf{x} \in \mathbb{R}^3 \mid x_1^2 + x_2^2 + x_3^2 \leq 1\} .$$

Remark 7.3 In the n-dimensional case $\mathscr{H} \cong \mathbb{C}^n$, this construction allows us to identify pure states as rank-one projectors in $\mathscr{B}(\mathscr{H})$. However, they will only be a closed portion of the $2(n - 1)$-dimensional unit sphere in $\mathbb{R}^{2n-1}$.

The tensor field Λ in the coordinates x_1, x_2, x_3 reads:

$$\Lambda = \epsilon_{ijk} x_i \frac{\partial}{\partial x_j} \wedge \frac{\partial}{\partial x_k} , \tag{7.30}$$

while the symmetric tensor field $\mathscr{R}$ is given by:

$$\mathscr{R} = \delta_{jk} \frac{\partial}{\partial x_j} \otimes \frac{\partial}{\partial x_k} - x_j x_k \frac{\partial}{\partial x_j} \otimes \frac{\partial}{\partial x_k} .$$

Remark 7.4 (On the Bivector Λ) The choice of the bivector Λ requires a comment. If we identify $\mathbb{R}^3$ with the dual of the Lie algebra of $SU(2)$, we can consider $\mathbf{x} = (x_1, x_2, x_3)$ as the linear functions on the dual of the Lie algebra $\mathfrak{su}(2)$ of $SU(2)$. Therefore the Lie bracket of $\mathfrak{su}(2)$ induces a Poisson bracket on $\mathfrak{su}(2)^*$

whose Poisson tensor is given by Λ. Notice that $SU(2)$ is the subgroup of unitary operators of determinant one of the group of unitary operators of $\mathcal{H} = \mathbb{C}^2$.

An alternative way of deriving Λ is to consider the projection map $S^3 \to S^2$ related with the momentum map associated with the symplectic action of the unitary group on the Hilbert space $\mathcal{H}$. Such map $\mu \colon \mathcal{H} \to \mathfrak{su}(2)^*$ provides a symplectic realization of the Poisson manifold $\mathfrak{su}(2)^*$.

In this context, observables correspond to affine functions on $\mathscr{S}$, that is, $f_a = a^j x_j + a_0, a_0, a_j \in \mathbb{R}$. Consequently, the Hamiltonian vector fields $X_a = \Lambda(df_a, \cdot)$, and the gradient-like vector fields $Y_a = \mathscr{R}(df_a, \cdot)$ are given by:

$$X_a = \epsilon_{jkl} a^j x_k \frac{\partial}{\partial x_l}, \qquad Y_a = a^j \frac{\partial}{\partial x_j} - a^k x_k \Delta,$$

with $\Delta = x_j \partial/\partial x_j$ the dilation vector field on $\mathbb{R}^3$. Lie algebra generated by the family of vector fields X_f, Y_f is the Lie algebra $SL(2, \mathbb{C})$.

It is now possible to construct a Lie-Jordan algebra (see for instance [6, 7, 14]) with commutative Jordan product $\circ$ and Lie product $\{\cdot, \cdot\}$ on the space of observables (affine functions) out of the tensors $\mathscr{R}$ and Λ. Such algebra is defined by:

$$x_j \circ x_k = \mathscr{R}(dx_j, dx_k) + x_j x_k, \qquad \{x_j, x_k\} = \Lambda(dx_j, dx_k).$$

Then we find:

$$x_j \circ x_j = 1, \qquad x_j \circ x_k = 0, \quad \forall j \neq k.$$

Combining the Jordan product and the Lie product we can define:

$$x_j \star x_k = x_j \circ x_k + i\{x_j, x_k\}$$

and we get:

$$x_j \circ x_k = \frac{1}{2}(x_j \star x_k + x_k \star x_j), \qquad \{x_j, x_k\} = -\frac{i}{2}(x_j \star x_k - x_k \star x_j)$$

The involution * will be complex conjugation and we get a C^*-algebra which can be used either to go back to the Hilbert space via de GNS construction or to go back to the Heisenberg picture if we realise the algebra in terms of operators.

Let us remark that as our algebras are described by means of tensor fields, it is evident that the particular coordinate system we use to describe the ball does not play any role. The convexity structure may well become hidden. For instance, parametrising Bloch's ball with spherical coordinates (r, θ, φ), the relevant tensor fields would be:

$$\mathscr{R} = (1 - r^2)\frac{\partial}{\partial r} \otimes \frac{\partial}{\partial r} + \frac{1}{r^2}\frac{\partial}{\partial \theta} \otimes \frac{\partial}{\partial \theta} + \frac{1}{r^2 \sin^2 \theta}\frac{\partial}{\partial \varphi} \otimes \frac{\partial}{\partial \varphi},$$

and

$$\Lambda = \frac{1}{r \sin \theta} \frac{\partial}{\partial \theta} \wedge \frac{\partial}{\partial \varphi}.$$

It is now clear by inspection that Hamiltonian vector fields and gradient vector fields are tangent to the sphere of pure states $S^2 = \{r = 1\}$. The interior of the ball is an orbit of the group $SL(2, \mathbb{C})$ and it is generated by the functions $r \cos \theta$, $r \sin \theta \sin \varphi$ and $r \sin \theta \cos \varphi$ by means of $\mathscr{R}$ and Λ.

To describe decoherence one needs vector fields which are generators of semigroups so that they will be directed vector fields not vanishing on the sphere of pure states.

7.2.1.2 Open Quantum Systems: the GKLS Equation

Let us consider the Kossakowski-Lindblad equation (see for instance [6] and references therein):

$$\frac{d}{dt}\rho = L(\rho),$$

with initial data $\rho(0) = \rho_0$ and,

$$L(\rho) = -i[H, \rho] + \frac{1}{2} \sum_j ([V_j \rho, V_j^\dagger] + [V_j, \rho V_j^\dagger])$$

$$= -i[H, \rho] - \frac{1}{2} \sum_j [V_j^\dagger V_j, \rho]_+ + \sum_j V_j \rho V_j^\dagger,$$

say with, $\operatorname{Tr} V_j = 0$, and $\operatorname{Tr}(V_j^\dagger V_k) = 0$ if $j \neq k$. We see immediately that the equations of motion split into three terms:

1. Hamiltonian term: $-i[H, \rho]$
2. Symmetric term (or gradient) : $-\frac{1}{2} \sum_j [V_j^\dagger V_j, \rho]_+$
3. Kraus term (or jump vector field): $\sum_j V_j \rho V_j^\dagger$.

It is possible to associate a vector field with this equation of motion [6, 7]. It turns out that the one associated with the Kraus term Z, is a nonlinear vector field, similar to the nonlinear vector field Y, associated with the symmetric tensor, the gradient vector field. The nonlinearity pops up because the two maps are not trace preserving therefore we have to introduce a denominator for the map to transform states into states. The "miracle" of the Kossakowski-Lindblad form of the equation is that the two nonlinearities cancel each other so that the resulting vector field is actually linear [6, 7].

Example 7.1 (The Phase-Damping of a q-Bit) Consider now:

$$L(\rho) = -\gamma (\rho - \sigma_3 \rho \sigma_3),$$

we find the vector field:

$$Z_L = -2\gamma \left(x_1 \frac{\partial}{\partial x_1} + x_2 \frac{\partial}{\partial x_2} \right)$$

which allows to visualise immediately the evolution.

7.3 Composition of Systems

As we mentioned in the introductory remarks, the composition of two systems A, B in the Dirac-Schrödinger picture is simply the tensor product $\mathscr{H}_A \otimes \mathscr{H}_B = \mathscr{H}_{AB}$. If our starting input is the complex projective space $P(\mathscr{H})$, we cannot consider the Cartesian product $P(\mathscr{H}_A) \times P(\mathscr{H}_B)$ because this would not contain all the information of the composite system, it would not contain what Schrödinger called the principal characteristic of quantum mechanics: the entangled states. According to our general procedure, we should associated with the composite system the complex projective space related to $\mathscr{H}_A \otimes \mathscr{H}_B$. It is easy to visualise the situation in the case of the qubit. Here the complex projective space is S^2, for two qubits we would have $S^2 \times S^2$. However if we take correctly the tensor product $\mathbb{C}^2 \otimes \mathbb{C}^2$ and then the associated complex projective space, we would get $P(\mathbb{C}^2 \otimes \mathbb{C}^2) = \mathbb{CP}^3$ which is six-dimensional and not four-dimensional as $S^2 \times S^2$. The additional states account for the entangled states, while the immersion of $S^2 \times S^2$ into $\mathbb{CP}^3$ would give the space of separable states.

A more intrinsic way would be to consider the tensor product $\mathscr{A}_A \otimes \mathscr{A}_B = \mathscr{A}_{AB}$ of the C^*-algebras $\mathscr{A}_A$ and $\mathscr{A}_B$ of expectation value functions on the Kähler manifolds of the physical subsystems, use the GNS construction to build a Hilbert space on which the chosen completion $\overline{\mathscr{A}_A \otimes \mathscr{A}_B}$ would have an irreducible representation, and the associated complex projective space should be considered to represent the composition of the two systems. Having the space describing the composite system we could proceed as usual.

7.3.1 Decomposing a System

Given the C^*-algebra $\mathscr{A}_{AB}$ of the total system we may now look for the two C^*-algebras, say $\mathscr{A}_A$ and $\mathscr{A}_B$, of the original components as subalgebras of the total C^*-algebra. We would ask of the subalgebras that they have in common only the identity and they commute with each other. Moreover we require that $\mathscr{A}_A \otimes \mathscr{A}_B$, after completion, be isomorphic with the total algebra.

To recover the states of the two subsystems we may define two projections, say: $\pi_A \colon \mathscr{S}_{AB} \to \mathscr{S}_A, \pi_A(\rho) = \rho_A, \rho_A(a) = \rho(a \otimes 1_B)$, and $\pi_B \colon \mathscr{S}_{AB} \to \mathscr{S}_B$, $\pi_B(\rho) = \rho_B, \rho_B(b) = \rho(1_A \otimes b)$, for all $a \in \mathscr{A}_A, b \in \mathscr{A}_B, \rho \in \mathscr{S}_{AB}$.

We find that $\rho_{AB} \neq \rho_A \otimes \rho_B$. Indeed, the quantity $\mathrm{Tr}\,(\rho_{AB} - \rho_A \otimes \rho_B)^k$ for every k, say integer, would provide possible measures of entanglement.

As a matter of fact both ρ_A and ρ_B are no more elements of the complex projective space associated to the two subsystems. They turn out to be, by construction, non-negative, Hermitian and normalised linear functionals, each one for the total C^*-algebra, that is, they are mixed states.

If we consider a unitary evolution on the composite system, say $U\rho U^\dagger$, we could consider, for any trajectory $U(t)\rho_0 U(t)^\dagger$, the projection on the subsystem $\mathscr{A}$, say:

$$\rho_A(t)(a) = (U(t)\rho_0 U(t)^\dagger)(a \otimes 1_B) = \rho_0(U(t)^\dagger(a \otimes 1_B)U(t)).$$

If ρ_0 is a separable pure state, it will project onto a pure state onto the subsystem. However, as time goes by, $\rho(t)$ will not be separable anymore and we get an evolution of a mixed state for the subsystem out of the evolution of a pure state for the total system. By letting the separable state ρ_0 vary by changing the second factor in $\mathscr{A}_B$ while preserving the first factor in $\mathscr{A}_A$, we would get an evolution for the projection on the system $\mathscr{A}_A$ which originates from the same initial point but would evolve with different trajectories, each one depending on the second factor.

When is it possible to describe the projected evolution by means of a vector field? This means that the projected trajectories would be described by a semigroup because the evolution would change the rank. The answer to this question was provided by A. Kossakowski and further formalised by Gorini, Kossakowski, Sudarshan and Linbland [15, 23]. The trajectories would be solutions of the Kossakowski-Lindblad master equation.

7.4 Conclusions and Discussion

The geometric description of mechanical systems based on the Kähler geometry of the space of pure states of a closed quantum system is proposed as an alternative picture of Quantum Mechanics. The composition of systems is also briefly discussed in this setting.

The tensorial description of Quantum Mechanics would allow for generic nonlinear transformations, hopefully more flexible to deal with nonlinearities, like entanglement, entropies and so on. Thus, the geometrical-tensorial description allows to recover as a covariance group of our description the full diffeomorphism group (similarly to General Relativity).

To illustrate the various aspects of the theory we study finite-dimensional systems, with a particular focus on the qubit example. It is shown that in the carrier space of the theory there are Hamiltonian and gradient vector fields X_a and Y_b generating the action of the Lie group $SL(\mathscr{H})$. This action may be extended to the closed convex body $\mathscr{S}$ of all quantum states. From the point of view of the affine ambient space $\mathfrak{T}_1$ of Hermitian operators with trace equal to 1 in which $\mathscr{S}$ naturally sits, we find that this action has, again, an infinitesimal description in terms of Hamiltonian and gradient-like vector fields closing on a realization of the Lie algebra $\mathfrak{sl}(\mathscr{H})$. Moreover, from the perspective of the evolution, to describe semigroups we have to introduce Kraus vector fields on $\mathfrak{T}_1$. Having described the dynamics in terms of vector fields will provide a framework to describe non-Markovian dynamics. States in the "bulk" may have as "initial conditions" pure, extremal states. The evolution would be described by a family of semigroups associated with higher order vector fields.

Acknowledgements The authors acknowledge financial support from the Spanish Ministry of Economy and Competitiveness, through the Severo Ochoa Programme for Centres of Excellence in RD (SEV-2015/0554). AI would like to thank partial support provided by the MINECO research project MTM2017-84098-P and QUITEMAD+, S2013/ICE-2801. GM would like to thank the support provided by the Santander/UC3M Excellence Chair Programme.

References

1. R. Abraham, J.E. Marsden, T. Ratiu, *Manifolds, Tensor Analysis, and Applications*, 3rd edn. (Springer, New York, 2012)
2. A. Ashtekar, T.A. Schilling. Geometrical formulation of quantum mechanics, in *On Einstein's Path: Essays in Honor of Engelbert Schucking* (Springer, New York, 1999), p. 42
3. S. Bochner, Curvature in Hermitian metric. Bull. Am. Math. Soc. **52**, 177–195 (1947)
4. M. Born, P. Jordan, Zur Quantenmechanik. Z. Phys. **34**(1), 858 (1925); M. Born, W. Heisenberg, P. Jordan, Zur Quantenmechanik. II. Z. Phys. **35**(8–9), 557 (1926)
5. F.M. Ciaglia, F. Di Cosmo, A. Ibort, G. Marmo, Dynamical aspects in the quantizer-dequantizer formalism. Ann. Phys. **385**, 769–781 (2017)
6. F.M. Ciaglia, F. Di Cosmo, A. Ibort, M. Laudato, G. Marmo, Dynamical vector fields on the manifold of quantum states. Open. Syst. Inf. Dyn. **24**(3), 1740003, 38 pp. (2017)
7. F.M. Ciaglia, F. Di Cosmo, M. Laudato, G. Marmo, Differential calculus on manifolds with boundary: applications. Int. J. Geom. Meth. Mod. Phys. **4**(8), 1740003, 39 pp. (2017)
8. R. Cirelli, A. Manià, L. Pizzocchero, Quantum mechanics as an infinite-dimensional Hamiltonian system with uncertainty structure: part I. J. Math. Phys. **31**, 2891–2897 (1990); *ibid.*, Quantum mechanics as an infinite-dimensional Hamiltonian system with uncertainty structure: part II. J. Math. Phys. **31**, 2898–2903 (1990)
9. A. Connes, *Noncommutative Geometry* (Academic, San Diego, 1994)
10. P.A.M. Dirac, *The Principles of Quantum Mechanics*, No. 27 (Oxford University Press, London, 1981)
11. E. Ercolessi, G. Marmo, G. Morandi, From the equations of motion to the canonical commutation relations. Riv. Nuovo Cimento Soc. Ital. Fis. **33**, 401–590 (2010)
12. G. Esposito, G. Marmo, G. Sudarshan, *From Classical to Quantum Mechanics* (Cambridge University Press, Cambridge, 2004)

13. P. Facchi, R. Kulkarni, V.I. Man'ko, G. Marmo, E.C.G. Sudarshan, F. Ventriglia, Classical and quantum Fisher information in the geometrical formulation of quantum mechanics. Phys. Lett. A **374**(48), 4801–4803 (2010)
14. F. Falceto, L. Ferro, A. Ibort, G. Marmo, Reduction of Lie-Jordan Banach algebras and quantum states. J. Phys. A Math. Theor. **46**(1), 015201 (2012)
15. V. Gorini, A. Kossakowski, E.C.G. Sudarshan, Completely positive dynamical semigroups of N-level systems. J. Math. Phys. **17**(5), 821–825 (1976)
16. J. Grabowski, M. Kuś, G. Marmo, Geometry of quantum systems: density states and entanglement. J. Phys. A Math. Gen. **38**(47), 10217–10244 (2005)
17. N.S. Hawley, Constant holomorphic curvature. Canad. J. Math. **5**, 53–56 (1953)
18. A. Ibort, V.I. Man'ko, G. Marmo, A. Simoni, F. Ventriglia, An introduction to the tomographic picture of quantum mechanics. Phys. Scr. **79**(6), 065013 (2009)
19. J. Igusa, On the structure of certain class of Kähler manifolds. Am. J. Math. **76**, 669–678 (1954)
20. P. Jordan, J. von Neumann, E. Wigner, On an algebraic generalization of the quantum mechanical formalism. Ann. Math. **35**(1), 29–64 (1934)
21. T.W.B. Kibble, Geometrization of quantum mechanics. Commun. Math. Phys. **65**, 189–201 (1979)
22. S. Kobayashi, K. Nomizu, *Foundations of Differential Geometry*, vol II (Wiley, New York, 1969)
23. G. Lindblad, On the generators of quantum dynamical semigroups. Commun. Math. Phys. **48**, 119–130 (1976)
24. G. Morandi, C. Ferrario, G. Lo Vecchio, G. Marmo, C. Rubano, The inverse problem in the calculus of variations and the geometry of the tangent bundle. Phys. Rept. **188**, 147–284 (1990)
25. L. Nirenberg, A. Newlander, Complex analytic coordinates in almost complex manifolds. Ann. Math. **65**(3), 391–404 (1957)
26. M. Skulimowski, Geometric POV-measures, pseudo-Kählerian functions and time, in *Topics in Mathematical Physics, General Relativity and Cosmology in Honor of Jerzy Plebanski*, Proceedings of the 2002 International Conference, Cinvestav Mexico City, 17–20 September 2002 (World Scientific, Hackensack, 2006), p. 433
27. J. von Neumann, *Mathematische Grundlagen der Quantenmechanik* (Springer, Berlin, 1932)
28. H. Weyl, Quantenmechanik und Gruppentheorie. Z. Phys. **46**, 1–46 (1927)
29. A. Wintner, The unboundedness of quantum-mechanical matrices. Phys. Rev. **71**(10), 738 (1947)

LECTURE NOTES OF THE UNIONE MATEMATICA ITALIANA

Editor in Chief: Ciro Ciliberto and Susanna Terracini

Editorial Policy

1. The UMI Lecture Notes aim to report new developments in all areas of mathematics and their applications - quickly, informally and at a high level. Mathematical texts analysing new developments in modelling and numerical simulation are also welcome.

2. Manuscripts should be submitted to
   Redazione Lecture Notes U.M.I.
   umi@dm.unibo.it
   and possibly to one of the editors of the Board informing, in this case, the Redazione about the submission. In general, manuscripts will be sent out to external referees for evaluation. If a decision cannot yet be reached on the basis of the first 2 reports, further referees may be contacted. The author will be informed of this. A final decision to publish can be made only on the basis of the complete manuscript, however a refereeing process leading to a preliminary decision can be based on a prefinal or incomplete manuscript. The strict minimum amount of material that will be considered should include a detailed outline describing the planned contents of each chapter, a bibliography and several sample chapters.

3. Manuscripts should in general be submitted in English. Final manuscripts should contain at least 100 pages of mathematical text and should always include

   – a table of contents;
   – an informative introduction, with adequate motivation and perhaps some historical remarks: it should be accessible to a
     reader not intimately familiar with the topic treated;
   – a subject index: as a rule this is genuinely helpful for the reader.

4. For evaluation purposes, please submit manuscripts in electronic form, preferably as pdf- or zipped ps-files. Authors are asked, if their manuscript is accepted for publication, to use the LaTeX2e style files available from Springer's web-server at
   ftp://ftp.springer.de/pub/tex/latex/svmonot1/ for monographs
   and at
   ftp://ftp.springer.de/pub/tex/latex/svmultt1/ for multi-authored volumes

5. Authors receive a total of 50 free copies of their volume, but no royalties. They are entitled to a discount of 33.3% on the price of Springer books purchased for their personal use, if ordering directly from Springer.

6. Commitment to publish is made by letter of intent rather than by signing a formal contract. Springer-Verlag secures the copyright for each volume. Authors are free to reuse material contained in their LNM volumes in later publications: A brief written (or e-mail) request for formal permission is sufficient.

Printed in the United States
By Bookmasters